D0219720

Praise for *Feminist Futures*:

'Provides a rich perspective on the lived experiences and agencies of women. A highly creative endeavour that will be valuable to activists and academics.'

Leela Fernandes, author of *Transnational Feminism in the United States*

'A diverse and exciting tapestry of themes and authors.'

Lourdes Beneria, Cornell University

'A candid and hard-hitting agenda for feminist scholarship and activism in the South in the twenty-first century.'

Patricia Mohammed, University of the West Indies

'The women, culture and development approach that the authors embrace is more prescient and necessary than ever.'

Amrita Basu, Amherst College

'Readable and well written ... especially valuable in the classroom.'

Choice

'[A] valuable and often challenging volume, a winding river that yields nuggets of gold.'

Gender and Development

CLIMATE FUTURES

CLIMATE FUTURES

RE-IMAGINING GLOBAL CLIMATE JUSTICE

Edited by Kum-Kum Bhavnani, John Foran,
Priya A. Kurian, and Debashish Munshi

ZED

Climate Futures: Re-imagining Global Climate Justice was first published in 2019 by Zed Books Ltd, The Foundry, 17 Oval Way, London SE11 5RR, UK.

www.zedbooks.net

Editorial Copyright © Kum-Kum Bhavnani, John Foran, Priya A. Kurian, and Debashish Munshi 2019
Copyright in this collection © Zed Books Ltd 2019

The right of Kum-Kum Bhavnani, John Foran, Priya A. Kurian, and Debashish Munshi to be identified as the editors of this work has been asserted by them in accordance with the Copyright, Designs and Patents Act, 1988

Typeset in Plantin and Kievit by Swales & Willis Ltd, Exeter, Devon, UK
Index by Austin Gee
Cover design by Steve Leard
Cover photo © Vlad Sokhin, Panos Pictures

Printed and bound by CPI Group (UK) Ltd, Croydon CR0 4YY

A catalogue record for this book is available from the British Library

ISBN 978-1-78699-781-4 hb
ISBN 978-1-78699-782-1 pb
ISBN 978-1-78699-783-8 pdf
ISBN 978-1-78699-785-2 epub
ISBN 978-1-78699-784-5 mobi

CONTENTS

TABLES

ACKNOWLEDGMENTS

This book took its first steps on the shores of Lake Como at Bellagio in Italy in July 2015, when we four organized a symposium on climate justice that brought together scholars and activists from around the globe to reflect on ways to enact the quest for climate justice. We thank the Rockefeller Foundation for providing us with a generous grant to facilitate the travel of delegates, and for hosting the symposium at the Foundation's picturesque Bellagio Center. The unique and beautiful setting of the Center made possible the vigorous and productive deliberations at the symposium. Thanks also to all our contributors, those who were at Bellagio as well as those who joined the journey later, for shaping the book, engaging in ongoing conversations, responding to endless questions, and revising their chapters as the book developed.

We are grateful to our home institutions – the University of California at Santa Barbara, USA, and the University of Waikato, Hamilton, New Zealand – for their support. In particular, Debashish and Priya thank the Waikato Management School and the Faculty of Arts and Social Sciences for on-going research support including academic visitor grants that made possible visits by Kum-Kum and John to the University of Waikato to work on this collaborative project.

In a critical period when university presses often drop the ball on the issues of the day (or century), we are grateful for the existence and support of Zed Books, both for its professionalism and for the reach of its message to diverse audiences around the world.

Finally, we share our deepest appreciation for the behind-the-scenes work done by our supremely competent, energetic, and efficient research assistants, Megan Smith in Hamilton, New Zealand, and Riley Hubbell in Santa Barbara, USA.

This book is dedicated:

by Kum-Kum, to her children: Cerina and Amal;

by John, to this generation of humans, to whom it falls to co-create the just climate future that the next seven are owed by us;

by Priya and Debashish, to their children: Akanksha and Alya, whose passion and commitment to justice and the environment represent the hope of the younger generation for a better future.

ABOUT THE CONTRIBUTORS

Anjali Appadurai is a climate communications specialist with West Coast Environmental Law in British Columbia, Canada. A human ecologist by training, she has worked on issues of environmental justice and legal reform on social, economic, and environmental aspects of climate change.

Seema Arora-Jonsson is professor at the Swedish Agricultural University. She works on climate politics through a feminist lens, especially in its connections across the Global North and South and with questions of environmental governance and development. Her recent work includes "Carbon and Cash: The Making of New Global Citizenship," *Antipode* (2016), "Blind Spots in Environmental Policy-Making: How Beliefs about Science and Development May Jeopardize Environmental Solutions," *World Development Perspectives* (2017), and "Does Resilience Have a Culture," *Ecological Economics* (2016).

Walter F. Baber is professor of public policy at California State University, Long Beach. He has received a number of Fulbright Awards including the Fulbright Distinguished Chair of Environmental Policy at the Politecnico di Torino (2009), the Fulbright visiting professor of political science at the Diplomatic Academy of Vienna (2016), and the Fulbright distinguished chair of public international law at the Raoul Wallenberg Institute of Human Rights at Lund University (2017/18).

Robert V. Bartlett is the Gund professor of liberal arts at the University of Vermont, USA. He has published many scholarly articles and book chapters and is an author or editor of ten books, most recently (with Walter F. Baber) *Consensus and Global Environmental Governance: Deliberative Democracy in Nature's Regime* (2015).

Nnimmo Bassey is an internationally known environmental activist, author, and poet. One of *Time* magazine's Heroes of the Environment in 2009 and co-winner of the Right Livelihood Award (2010), he is a former chair of Friends of the Earth International and executive director of Environmental Rights Action. He is based in Nigeria.

Kum-Kum Bhavnani is distinguished professor of sociology at the University of California at Santa Barbara, USA. She works in Third World development, feminist and cultural studies. She has published a number of books, edited collections and articles including *Talking Politics* (1991), *Feminism and "Race"* (2001) and *Feminist Futures* (2nd edition, Zed Books; 2016). She disseminates much of her research through independent documentary films, including *The Shape of Water* (2006), *Nothing Like Chocolate* (2012), *Lutah* (2014), and *We Are Galapagos* (2018).

Patrick Bond is professor of political economy at the University of the Witwatersrand School of Governance in Johannesburg, South Africa. His doctorate was under David Harvey's supervision at Johns Hopkins University, USA, and his books include *Politics of Climate Justice: Paralysis Above, Movement Below* (2011), *Elite Transition: From Apartheid to Neoliberalism in South Africa* (2014), and *BRICS: An Anti-Capitalist Critique* (2016).

Anabela Carvalho is associate professor in the Department of Communication Sciences at the University of Minho, Portugal. Her research focuses on various forms of environment, science, and political communication with a particular emphasis on climate change. Her publications include *Communicating Climate Change: Discourses, Mediations and Perceptions* (2008), *Citizen Voices: Enacting Public Participation in Science and Environment Communication* (with L. Phillips and J. Doyle; 2012), *Climate Change Politics: Communication and Public Engagement* (with T.R. Peterson; 2012).

Dipesh Chakrabarty is the Lawrence A. Kimpton distinguished service professor of history at the University of Chicago, USA. Recognized as one of the foremost scholars of postcolonial theory and subaltern studies, he has written several influential books including the much cited *Provincializing Europe: Postcolonial Thought and Historical Difference* (2000) and *Habitations of Modernity: Essays in the Wake of Subaltern Studies* (2002).

Nigel Clark is professor of human geography at Lancaster University, UK. He is the author of *Inhuman Nature: Sociable Life on a Dynamic Planet* (2011) and co-editor (with Kathryn Yusoff) of a recent *Theory, Culture and Society* special issue on "Geosocial Formations and the Anthropocene" (2017). He is working on a book (with Bronislaw Szerszynski) about planetary thinking for the social sciences.

John Foran is professor of sociology and environmental studies at the University of California, Santa Barbara. He is a passionate scholar-activist with the global climate justice movement. Among his books are *Taking*

Power: On the Origins of Revolutions in the Third World (2005), and *2015: Year Zero for Climate Change* (forthcoming). His work can be found at www. resilience.org and www.climatejusticeproject.org.

Greta Gaard is professor of English and founding coordinator of the Sustainable Justice Minor at the University of Wisconsin-River Falls, USA. Author or editor of six books and over 100 articles and international presentations, Gaard's most recent volume, *Critical Ecofeminism* was published May 2017. Her creative nonfiction eco-memoir, *The Nature of Home*, has been translated into Chinese and Portuguese.

Sumetee Pahwa Gajjar is former lead practice with the Indian Institute for Human Settlements, Bangalore, India. She researches and writes on urban policy, environmental sustainability, climate change, and nature-based solutions.

Corrie Grosse is assistant professor of environmental studies at College of Saint Benedict and Saint John's University, USA, where she teaches energy and society, gender and environment, and social responses to climate change. She specializes in the intersection of energy extraction, climate justice, and grassroots organizing. Her current research examines how communities in Idaho and California work together to resist hydraulic fracturing and tar sands. To learn more, visit her website (www.corriegrosse.com) or e-mail her at cgrosse001@csbsju.edu.

Yasmin Gunaratnam is a reader in sociology at Goldsmiths (University of London), UK. Her books include *Researching Race and Ethnicity: Methods, Knowledge and Power* (2003), *Death and the Migrant* (2013), and *Go Home? The Politics of Immigration Controversies* (2017). She has edited nine collections and is on the editorial collectives of *Feminist Review* and *Media Diversified*.

Ken Hiltner is professor of the environmental humanities and director of the Environmental Humanities Initiative at the University of California at Santa Barbara, USA. He has written and edited several books including *Milton and Ecology* (2003), *Renaissance Ecology: Imagining Eden in Milton's England* (2008), *What Else Is Pastoral? Renaissance Literature and the Environment* (2011), *Environmental Criticism for the Twenty-first Century* (2011), *Ecocriticism: The Essential Reader* (2014), and *Writing a New Environmental Era: Moving Forward to Nature* (forthcoming).

Brad Hornick is a scholar and activist pursuing a PhD in sociology at Simon Fraser University, Canada. He is an integral part of System Change Not Climate Change – an Ecosocialist Network.

Saleemul Huq is director of the International Centre for Climate Change and Development at Independent University, Bangladesh and senior fellow, International Institute for Environment and Development, London. He has worked extensively on the inter-linkages between climate change and sustainable development from the perspective of developing countries, with special emphasis on least developed countries. He has published numerous articles in scientific and popular journals, and was a lead author of the Third Assessment Report of the Intergovernmental Panel on Climate Change.

Lagipoiva Cherelle Jackson, an independent journalist based in Samoa, is a specialist on Pacific climate change issues and has played an active role in global climate change negotiations. She is a founder of the Pacific Alliance of Development Journalists, edits *Environment Weekly*, and is part of several research projects on climate change in the Pacific Islands.

Garima Jain is a senior consultant at the Indian Institute for Human Settlements, Bangalore, India. She works in disaster risk reduction, climate change adaptation, and human development from the perspectives of urban planning practices and policies. She has been part of the Secretariat for the Sustainable Development Goals agenda for cities (SDG 11), and has led projects funded by UNDP and the Rockefeller Foundation. She also serves on the Urban Planning Advisory Group at the United Nations.

Priya A. Kurian is professor of political science and public policy at the University of Waikato, New Zealand. Her current research, funded by national and international grants, focuses on climate change politics and policy and sustainable development. Her scholarly works include *Engendering the Environment? Gender in the World Bank's Environmental Policies* (2000), *Feminist Futures: Re-imagining Women, Culture, and Development* (Zed Books, 2003; 2016) and articles in *Nature Climate Change*, *Public Understanding of Science*, *Citizenship Studies*, and *Policy and Politics*.

Sherilyn MacGregor is a reader in environmental politics at the University of Manchester. As a specialist in the interdisciplinary field of gender and environmental politics, her research looks at environmental (un)sustainability, gender (in)equality, and theories and practices of citizenship. She is the editor of the *Routledge Handbook of Gender and Environment* (2017) and served as an editor of *Environmental Politics* until early 2017.

Andreas Malm teaches human ecology at Lund University, Sweden. His most recent book is *The Progress of This Storm: Nature and Society in a Warming World* (2018).

Bill McKibben is an environmentalist, author, and academic who has written extensively on the impact of global warming. In 2008, he co-founded what became the global climate justice organization, 350.org, with a small group of students at Middlebury College, USA. His books include *The End of Nature* (1987), *Eaarth: Making a Life on a Tough New Planet* (2010), *The Global Warming Reader* (2011), *Oil and Honey: The Education of an Unlikely Activist* (2013), and the novel *Radio Free Vermont: A Fable of Resistance* (2017).

Kavya Michael is an associate fellow at the Energy and Resources Institute in New Delhi, India, with a research background in human ecology, the political economy of the environment, migration, as well as environmental inequalities. She has written extensively on the multiple intersections of climate change/environmental hazards, urban inequality, and development in Indian cities. Her work is oriented strongly within the urban climate justice paradigm and has emphasized the need for bringing a climate justice lens to cities.

Sandra L. Morrison is an associate professor of Māori and Indigenous studies at the University of Waikato, New Zealand. She is currently the Vision Mātauranga Leader and a member of the Senior Leadership Team of the Deep South National Science Challenge set up by the New Zealand government to prepare the country to adapt to and manage the risks of climate change.

Meraz Mostafa has worked as a research officer at the International Centre for Climate Change and Development (ICCCAD) in Dhaka, Bangladesh for the last three years. He works under the migration and food security programs at the center, and manages the content of a special monthly magazine titled the *Climate Tribune* on climate change issues relevant to Bangladesh and beyond.

Debashish Munshi is professor of management communication at the University of Waikato, New Zealand. His current research focuses on public engagement on complex and contentious issues. He was co-principal investigator of a study on "Sustainable Citizenship: Transforming Public Engagement on New and Emerging Technologies" funded by a Marsden Grant of the Royal Society of New Zealand. He is currently co-leader on a National Science Challenge project on "Centring Culture in Public Engagement on Climate Change Adaptation" in New Zealand.

Sunita Narain, an environmentalist and writer, is the director-general of the Centre for Science and Environment, New Delhi, India, and editor of the fortnightly *Down To Earth*. She plays an active role in policy formulation on issues of environment and development in India and globally. She has co-authored influential publications on India's environment and conducted extensive research on environmental governance and management. In 2016, she was featured in *Time* magazine's list of 100 Most Influential People in the World.

Naznin Nasir is a PhD student in the Department of Geography at Durham University, UK. She has worked for the International Centre for Climate Change and Development, Bangladesh, on a number of projects on migration, climate change adaptation, and the political economy of clean energy development. She has also had a career as a broadcast journalist.

Jess Pasisi is a Niuean-New Zealand PhD student studying at the University of Waikato, New Zealand. She is researching the cultural practices, knowledge, and lived experiences of Niuean people in relation to climate change. As the complexities of climate change have ever-increasing bearing on how the Niuean people live day to day, her thesis captures some of the unique knowledge of these people that is integral to how these communities adapt to and mitigate the impacts of climate change.

David N. Pellow is the Dehlsen chair and professor of environmental studies and director of the Global Environmental Justice Project at the University of California at Santa Barbara, USA. His teaching and research focus on environmental and ecological justice in the US and globally. He has served on the boards of directors for the Center for Urban Transformation, Greenpeace USA, and International Rivers.

Anna Pérez Català is the campaigns director (Spain) for Climate Tracker, a global collective of young reporters and influential voices for action on climate change. She is an environmental scientist specializing in climate change and development.

Jonathon Porritt, founder director of Forum for the Future, is a leading environmentalist and author of several highly-acclaimed books including *Capitalism as if the World Matters* (2005) and *The World We Made* (2013). He was the inaugural chair of the Sustainable Development Commission in the UK and is a former director of Friends of the Earth.

M. Feisal Rahman is a research coordinator at the International Centre for Climate Change and Development, Bangladesh. A trained environmental engineer, he has been involved with numerous projects on adaptation technology, urban climate change resilience, and capacity development in management and allocation of natural resources.

Kim Stanley Robinson is an American science fiction writer, the author of around 20 books. He has visited the Antarctic twice, courtesy of the US National Science Foundation. In 2008 he was named a "Hero of the Environment" by *Time* magazine, and he works with the Sierra Nevada Research Institute, the Clarion Writers' Workshop, and UC San Diego's Arthur C. Clarke Center for Human Imagination. His work has been translated into 25 languages.

Maria Alejandra (Majandra) Rodriguez Acha is a Peruvian activist and educator. She has a BA in sociology/anthropology and sustainable development (Swarthmore College). Her work focuses on intersectionality, feminisms, decolonialism, and epistemologies of the South, and the building of alternatives to the environmental crisis. She is on the board of directors at FRIDA | Young Feminist Fund, an advisor on the Next Generation Climate Board at Global Greengrants Fund, and co-founder of TierrActiva Perú, a network for "system change, not climate change."

Chandni Singh is a researcher and faculty member at the Indian Institute for Human Settlements in Bangalore, India. She works on issues of climate change adaptation, livelihood vulnerability, and migration with a regional focus on South Asia.

Pablo Solón is an environmental activist who served as Bolivia's ambassador to the United Nations from 2009 to 2011. Thereafter, he was executive director of Focus on the Global South, an international activist think tank based in Bangkok (2012 to 2015). A former winner of the International Human Rights Award, he remains active in Indigenous movements, workers' unions, and cultural organizations in Bolivia.

Sangion Appiee Tiu is the director for the Research and Conservation Foundation (RCF), a not-for-profit conservation organization based in Papua New Guinea. Prior to that, she completed her PhD at the University of Waikato, New Zealand. Her research interests are in the areas of traditional ecological knowledge for natural resource management, environment and sustainability education, and bio-cultural education.

Rikard Warlenius is a journalist and social worker in Sweden and is currently a PhD candidate in human ecology at Lund University. He has served on the board of the Climate Change Association and is a member of the Left Party's Ecological Economic Program Committee.

Kyle Powys Whyte holds the Timnick Chair in the humanities and is a professor of philosophy and community sustainability at Michigan State University, USA. An enrolled member of the Citizen Potawatomi Nation, his work focuses on the problems and possibilities Indigenous peoples face regarding climate change, environmental justice, and food sovereignty. His articles have appeared in journals such as *Climatic Change, Daedalus, Hypatia, Synthese, Human Ecology, Journal of Global Ethics*, and *Journal of Agricultural and Environmental Ethics*.

Fengshi Wu is a senior lecturer at the Asia Institute of the University of Melbourne, Australia. Before moving to Australia, she worked at the Nanyang Technological University, Singapore, and at the Chinese University of Hong Kong. A specialist in environmental politics, Chinese politics and global governance, she has published in several international journals including the *International Studies Quarterly, China Journal, Global Policy*, the *International Journal of Voluntary and Nonprofit Organizations, Issues and Studies*, and the *Journal of Environmental Policy and Planning*.

Mohamed Hamdhaan Zuhair holds a Bachelor's degree in environmental science from James Cook University, Australia, and a Master's in management studies (management and sustainability) from the University of Waikato, New Zealand. His research interests include environmental impact assessment, climate change, environmental sustainability, and environmental decision making. He has worked in the environment sector in the Maldives for the past nine years, in the Environmental Protection Agency and most recently in a number of donor-funded projects.

ABBREVIATIONS

ABC	Afrikan Black Coalition
AIDS	Acquired Immunodeficiency Syndrome
AMRUT	Atal Mission for Rejuvenation and Urban Transformation
AOSIS	Alliance of Small Island States
BAU	business as usual
BCCT	Bangladesh Climate Change Trust
BDS	Boycott Divestment Sanctions
CBDR	common but differentiated responsibilities
CBIT	Capacity Building Initiative for Transparency
CCAP	Climate Change Adaptation Project
CCCAN	China Civil Climate Action Network
CDM	Clean Development Mechanisms
CERD	Committee on the Elimination of Racial Discrimination
CFC	chlorofluorocarbons
CJ	climate justice
CNKI	Chinese National Knowledge Infrastructure
COP	Conference of the Parties
COSATU	Congress of South African Trade Unions
CYCAN	China Youth Climate Action Network
DAPL	Dakota Access Pipeline
EIA	Environmental Impact Assessment
EJ	environmental justice
EJOLT	Environmental Justice Organizations, Liabilities and Trade
EPA	Environmental Protection Agency
EU	European Union
G-77	Group of 77 developing countries
GCF	Green Climate Fund
GDP	Gross Domestic Product
GEF	Global Environment Facility
GEI	Global Environmental Institute
GH	Greenovation Hub
GHG	greenhouse gas
Gt	gigatons
HIV	Human Immunodeficiency Virus
IMF	International Monetary Fund

INDC	Intended Nationally Determined Contributions
IPCC	Intergovernmental Panel on Climate Change
ISIL	Islamic State of Iraq and the Levant
ITUC	International Trade Union Confederation
JNNURM	Jawaharlal Nehru National Urban Renewal Mission
km	kilometre
KP	Kyoto Protocol
kph	kilometres per hour
kpl	kilometres per litre
LDC	Least Developed Countries
LGBT	lesbian, gay, bisexual, trans
LGBTQ+	lesbian, gay, bisexual, trans, queer/questioning, and others
LNG	Liquefied Natural Gas
LUCCC	LDCs Universities Consortium on Climate Capacity
m	metre
MfE	Ministry for the Environment
mi.	mile
MoEFCC	Ministry of Environment, Forests and Climate Change
mpg	miles per gallon
mph	miles per hour
MSL	Mean Sea Level
NAPCC	National Action Plan on Climate Change
NBA	Narmada Bachao Andolan (or Save the Narmada Movement)
NDC	Nationally Determined Contributions
NGO	non-governmental organization
NIMBY	not in my back yard
NITI	National Institution for Transforming India
NIWA	National Institute for Water and Atmospheric Research
NOW	Niue Ocean Wide
OECD	Organisation for Economic Co-operation and Development
PCCB	Paris Committee on Capacity Building
PIFS	Pacific Islands Forum Secretariat
PNG	Papua New Guinea
QFFL	Queer Food For Love
QTPOC	queer and trans* people of color
RCF	Research & Conservation Foundation
REDD	Reducing Emissions from Deforestation and Degradation
REDD+	Reducing Emissions from Deforestation and Forest Degradation Plus
RHC	Reindeer Herding Communities

Rinos	Republicans In Name Only
SAPCC	State Action Plans on Climate Change
SIDS	Small Island Developing States
SLAPPs	Strategic Lawsuits Against Public Participation
STEM	Science, Technology, Engineering, and Math
TEK	traditional ecological knowledges
UC	University of California
UK	United Kingdom
UKIP	United Kingdom Independence Party
UN	United Nations
UNDP	United Nations Development Programme
UNEP	United Nations Environment Programme
UNFCCC	United Nations Framework Convention on Climate Change
US	United States
WCS	Wildlife Conservation Society
WHO	World Health Organization
WRI	World Resources Institute
WTO	World Trade Organization

FOREWORD

Equity: the final frontier for an
effective climate change agreement

Sunita Narain

In 1992, when the world met to discuss an agreement on climate change, equity was a simple concept: sharing the global commons—the atmosphere in this case—equally among all. It did not provoke much anxiety, for there were no real claimants. However, this does not mean the concept was readily accepted. A small group of industrialized countries had burnt fossil fuels for 100 years and built up enormous wealth. This club had to decide what to do to cut emissions, and it claimed all countries were equally responsible for the problem.

Just as the climate convention was being finalized in 1991, a report, released by an influential Washington think tank, World Resources Institute (WRI), broke the news that India, China, and other developing countries were equally responsible for greenhouse gases. My colleague Anil Agarwal and I rebutted this argument in our 1991 publication, *Global Warming in an Unequal World: A Case of Environmental Colonialism.*

Based on our review of the methodology used by WRI to calculate the "net" emissions of each country, we made two main points.

One, the world needed to differentiate between the emissions of the poor—from subsistence paddy or animals—and that of the rich—from, say, cars. Survival emissions were not and could not be equivalent to luxury emissions.

Two, it was clear that managing a global common meant cooperation between countries. As a stray goat or cattle is likely to chew up saplings in the forest, any country could blow up the agreement if it emitted beyond what the atmosphere could absorb. Cooperation was only possible—and this is where our forests experience came in handy—if benefits were distributed equally. We then developed the concept of per capita entitlements—each nation's share of the atmosphere—and used the property rights of entitlement to set up rules of engagement that were fair and equitable. We said that countries using less than their share of the atmosphere could trade their unused quota and this would give them the incentive to invest in technologies that would not increase their emissions. But in all this, as we told climate negotiators, think of the local forest and learn that the issue of equity is not a luxury. It is a prerequisite.

This was the inconvenient truth.

In 1992, it was accepted that the occupied atmospheric space would need to be vacated to make room for the emerging world to grow because emissions are an outcome of economic growth. This acceptance recognized the principle of common but differentiated responsibilities (CBDR) in reducing emissions. A firewall was built to separate those countries that had to reduce emissions to make space for the rest of the world to grow. That year in Rio de Janeiro, the world was talking about drastic cuts of 20 percent below the 1990 levels to provide for growth as well as climate security. Even in that age of innocence, the negotiations were difficult and nasty. The US argued its lifestyle was non-negotiable and refused to accept any agreement specifying deep reductions. In 1997, the Kyoto Protocol set the first legal target for these countries much below what the world knew it needed to achieve.

Two and a half decades later, the idea of equity has become an even more inconvenient truth. By now there are more claimants for atmospheric space. Emerging countries have now emerged. China, which in 1990, with over a quarter of the world's population, was responsible for only 10 percent of annual emissions, contributed 27 percent by 2016. So, the fight over atmospheric space is now real. While the rich countries have not reduced emissions, the new growth countries have started emitting more. In 1990, the industrialized countries accounted for 70 percent of the global annual emissions. In 2016, they accounted for 43 percent, but this is not because they have vacated space. The new growth countries—China in particular—have only occupied what was available. Emission reductions proposed 20 years ago have still not been committed or adhered to. In fact, in most already industrialized countries, emissions have either stabilized or increased. In coal and extractive economies, like Canada and Australia, emissions have risen by 20 percent and 46 percent respectively.

The world has run out of atmospheric space and certainly of time. The question now is if the rich, who contributed to emissions in the past, will still take up an unfair share of this space based on their populations' overconsumption, or reduce both consumption and emissions? Or will the emerging countries be told to take over the burden? This is the big question, and an inconvenient one at that.

And this when climate change is not just a problem of the present but also of past contributions as well. The stock of greenhouse gases in the atmosphere has a long life. This means that any discussion on how the carbon cake will be divided, must take into account those gases emitted in the past and still present. So, while China accounts for 27 percent of the annual emissions, in cumulative terms (since 1950) it still accounts for only 11 percent of total emissions since the 1800s. Similarly, India contributes 6 percent to the annual global emissions, but is only responsible for 3 percent of the total stock. The rich countries, with less than a quarter of the world's population, are responsible for some 70 percent of this historical burden. This stock of gases is responsible for an average global temperature rise of 0.8°C and another 0.8°C in future, which is inevitable, as the gasses are already in the atmosphere.

The fact also is that temperature increase has a clear correlation with emissions. Think of it like a cake or a budget—what is eaten, what is taken, what remains. Crumbs or not.

The budgetary facts are as follows: According to the Intergovernmental Panel on Climate Change (IPCC), of the 2,900 billion metric tons of carbon dioxide that the world can emit if it wants to stay below 2°C, some 1,900 billion metric tons have been used up—this amount of carbon dioxide is already accumulated in the atmosphere. There are some 1,000 billion metric tons left, which can be emitted between now and 2100.

The already industrialized countries have overused their carbon quota. But more importantly, their lack of ambition means they will continue to surreptitiously appropriate even more of the budget.

The US, with just 5 percent of the world's population, for instance, has already used up some 21 percent of the total "used up" carbon budget—that is, it is responsible for 21 percent of the 1,900 billion metric tons emitted from 1850 to 2011. Between now and 2030, based on its rather meaningless commitment to reduce emissions, it will emit another 92 billion metric tons and take up another 8 percent. In this way, each country's emission reduction target reflects its intention to occupy global carbon space.

Based on the aggregation of the current emission targets of all countries, by 2030 some 80 percent of the total carbon budget available to the world to keep temperatures below 2°C will be used up.

This would be fine if all countries were at equal levels of development. Then nobody would require ecological space for economic growth beyond 2030. But this is hardly the case. India and almost all of Africa, even under the most aggressive plans for growth, would still be struggling to meet the basic needs of people beyond 2030. But by then the carbon budget would be all appropriated and gone.

The fact is that if temperature increase is capped at $1.5°C$, then the carbon budget—how much the world can emit to cap that temperature rise—is limited even further. In fact, at current rates of emissions this "budget" will be more or less exhausted by 2020. I ask again. If this is not injustice, what is?

"Equity" is an idea that is difficult to sell in a world distrustful of idealism and any talk of distributive justice. Even developing country negotiators do not really believe this form of climate-socialism can happen. They will tell you that the world is never going to give up space; that the world is too mean to give money or technology to poor nations for transition to low-carbon growth.

But it is also a fact that cooperation is not possible without fairness and equity. This is why I call equity the pre-requisite. If the US, Japan, and even China polluted in the past, then India and Africa will pollute in the future—unless there is clear agreement to limit everybody's emissions with a measure of fairness. So, let's understand that equity is about an effective agreement.

But, as I said, the idea of equity is inconvenient. There are also questions asked about how legitimate is this argument of equity, when it comes from large polluters like India.

My colleagues at the Centre for Science and Environment analyzed income distribution and income elasticity of emissions data to see if rich Indians emitted more or as much as rich country counterparts. The study (Centre for Science and Environment 2011) found that the per capita emissions of the richest 10 percent of India's population was the same or slightly less than the per capita emissions of America's poorest 10 percent and it was less than one-tenth the per capita emissions of America's richest 10 percent. In other words, the rich of India emitted less than the poorest American and there was no comparison between them and the rich of that world. If we accept this, then we accept to freeze inequity in the world forever.

It's time we stopped this kindergarten fight. Let us be clear that the world has to cut emissions drastically and fast. There must be limits on all countries. But these emission reductions will have to be based on the principle of equity so that the right to development is secured for all.

This is the most inconvenient of truths. But it is the truth.

This is the most inconvenient truth that Donald Trump wants to erase forever. When he announced on June 2, 2017 that his country would exit from the Paris Agreement, he not only rejected the urgency of climate change, but also the issue of equity. Trump hit out at developing countries saying they would unfairly benefit from the Paris Accord. "China will be able to increase these emissions by a staggering number of years—13. They can do whatever they want for 13 years. Not us. India makes its participation contingent on receiving billions and billions and billions of dollars in foreign aid from developed countries" (Trump 2017). The Trump doctrine of fairness has a new definition. The powerful can take it all and still call the shots.

This is why we must assert the imperative of equity once again. This time, in the post-Trump world. To date, the US has usually made the multilateral world change rules; reconfigure agreements, mostly to reduce it to the lowest common denominator, all to get its participation. Then when the world has a weak, worthless, and meaningless deal, it will walk out of this. All this while its powerful civil society and media will hammer on the point that the world needs to be accommodating and pragmatic. "Our Congress will not accept" is the refrain—essentially arguing that theirs is the only democracy in the world or certainly the only one that matters.

This happened in 1992, when in Rio, after much "accommodation," the agreement to combat climate change was whittled down; targets were removed, there was no agreed action. All this was done to bring the US on board. But it walked out. Then came the Kyoto Protocol—the first and only framework for action to reduce emissions. Here again, in December 1997, when climate change proponents Bill Clinton and Al Gore were in office, the agreement was reduced to nothingness— the compliance clause was removed; cheap emissions reduction added; loopholes included. All to bring the US on board. Once again, they rejected it.

Then came Barack Obama and his welcome commitment to climate change actions. But what did the US do? It has made the world completely rewrite the climate agreement so that, instead of targets based on science and contribution of each country, it is now based on voluntary action. Each country is allowed to set targets, based on what they can do and by when. It has led to weak action, which will not keep the planet's temperature below $2°C$ (forget the guardrail of $1.5°C$). This was done to please the Americans who said that they would never sign a global agreement which binds them to actions or targets. Paris fatally and fundamentally erased all historical responsibility of countries and reduced equity to sweet nothings. This was done because the US said that this was the redline—nothing on equitable rights to the common atmospheric space could be acceptable.

This is where we are today. We have Donald Trump, who openly denies climate change, at the helm in the US. A large majority stands with him. We have growing calls for protectionism in this already rich world—the UK's Brexit vote is also a testimony to this anger. It is the revenge of the rich, who did not get richer. It is the revenge of the educated well-off, who believe they are entitled to more and that this is being taken away from them by "others." This is also a time then that the already developed world, which has by now long exhausted whatever quota it had of the global atmospheric space, wants to burn more fossil fuels for its growth. It believes it is growth-deprived.

The key reason for all this is the fact that globalization has increased inequity. This is at the core of the problem today. This is also the core of climate change—ultimately, if emissions are linked to economic growth, then the question is how this growth will be shared between people and between nations. Economic and ecological globalization are about making rules that benefit people and the planet—not in ways that some get richer and not in ways that we blow up the climate.

This is what we need to work on in the post-Trump world. But this demands a change in the narrative. For too long, the two discussions on growth and climate change have been separated. For too long, we have been told that we cannot discuss issues of equitable growth or equitable allocation of the carbon budget. This is what needs to change.

Otherwise, we will be in denial. And the climate deniers will have the last laugh.

1 | THE FUTURE IS OURS TO SEEK: CHANGING THE INEVITABILITY OF CLIMATE CHAOS TO PROSPECTS OF HOPE AND JUSTICE

Debashish Munshi, Priya A. Kurian, John Foran, and Kum-Kum Bhavnani

The present is tense but it is within our hands to make sure that the future is not. As in grammar, it is the will to change that marks the shift in tenses. And to combat the forces of climate change bearing down on the planet we need to harness the power of the will to change. We are long past the time to sit back and accept that whatever will be will be. As the author and social activist Naomi Klein (2014a: 28) says, "[N]othing is inevitable. Nothing except that climate change changes everything. And for a very brief time, the nature of that change is up to us."

Around the time Klein's (2014a) landmark book *This Changes Everything* was stirring social media platforms and receiving widespread attention in the news media, another influential author, Margaret Atwood (2014), suggested that the term "climate change" was limiting and that we should refer to the "great big change we are all in the midst of" as "everything change." This is because, as she says, when there are climatic changes as, for example, the patterns of where it rains or where it doesn't, just about everything in life changes—from the foods we eat to the places we live in, to our relationships with human and non-human life and to nature in general. How drastic such changes can be is on constant display and brought home to us if not through direct experience then by the news and social media. The grim reality of climate change unfolding around us, for example, is captured in the extreme weather events around the world, wreaking havoc and setting new records in their wake. George Monbiot's (2017) words about Hurricane Harvey that hit Texas in August 2017, could just as well apply to Hurricanes Irma and Maria that devastated large swathes of the Caribbean a month later or indeed Cyclones Fehi and Gita that ravaged the South Pacific in 2018:

Hurricane Harvey offers a glimpse of a likely global future; a future whose average temperatures are as different from ours as ours are from those of the last ice age. It is a future in which emergency becomes the norm, and no state has the capacity to respond. It is a future in which, as a paper in the journal *Environmental Research Letters* notes, disasters like Houston's occur in some cities several times a year. It is a future that, for people in countries such as Bangladesh, has already arrived, almost unremarked on by the rich world's media. It is the act of not talking that makes this nightmare likely to materialize. (Monbiot 2017, citing Buchanan, Oppenheimer, and Kopp 2017)

Our volume wants to create possibilities for talking so that the nightmare does not materialize. At the same time, we editors have a relentless hope—also common to Klein's and Atwood's visions—for a radically different future that encompasses a flourishing and just world.

It was not only Klein's and Atwood's musings that got us thinking. It was also the stories of the countless women and men working tirelessly at the frontlines of climate change in every part of the world—from the sinking landmasses of the Southern Pacific and the drought-stricken stretches of Africa and West Asia to the flood-ravaged cities and villages of South Asia and South and Central America. Again, what was common to these stories was a sense of hope in the midst of tragedy, a phenomenon well-documented by Rebecca Solnit in her 2009 classic, *A Paradise Built in Hell: The Extraordinary Communities That Arise in Disaster.*

It is our confidence in the active power of hope that the four of us embarked on a journey to envision *Climate Futures* and to join with others to re-imagine global climate justice. We had already worked on a volume, *Feminist Futures* (Bhavnani, Foran, Kurian, and Munshi 2016), inspired as we were by the extraordinary ways in which subaltern Third World women and men resist the oppressive conditions in which they find themselves. We decided that what was needed was to find a way to bring people from around the world for conversations and debates around issues of climate justice. A grant from the Rockefeller Foundation allowed us to host a symposium at the Rockefeller Center in Bellagio, Italy, in July 2015. The goal was to see how climate justice could be collectively re-imagined by climate justice organizations, grassroots activists, public intellectuals, and scholars. Those intense Bellagio discussions led us to craft an open letter in December 2015, published in the *Huffington Post*, to Christiana Figueres, the then Executive Secretary to the United Nations Framework Convention on Climate Change

(UNFCCC), Laurent Fabius, the President of the 21st UN Conference of the Parties (COP 21), and the would-be negotiators of the world's future from 195 countries at the Paris COP 21. We asked that delegates "center the idea of *justice* in climate negotiations" and put "gender, indigenous rights, and resource distribution on the agenda so we may all enjoy a low-carbon, sustainable, equitable, and deeply democratic future" (Bhavnani 2015; see Foran 2016c on the limitations of the outcome of COP 21's "Paris Agreement").

This volume has emerged from the discussions we began in Bellagio, which continued with an enlarged community of scholars and activists. We took on the challenge of mapping strategies toward a better future for our planet, or at a minimum, the best-case planet that we can get to. Our aim has been to sharpen understandings of the social, political, and cultural aspects of environmental change, as climate change and climate justice cannot be thought about—or better futures imagined—without them. We do this by drawing on the energy of climate action groups, and their multifaceted responses to questions of justice, to envision planetary social movements that could help spark the necessary changes for sustainable economic, social, political, and environmental policies and practices.

Climate justice perspectives center the fact that the brunt of climate change falls hardest on the most poor and marginal peoples—peoples often trampled by the twin ravages of colonialism and capitalism, who demonstrate resilience despite these depredations. The rampant extraction of resources by imperial powers in colonized lands—and subsequently by local predator elites—left the lands in a state of continuing impoverishment, and with depleted levels of physical and economic resources that make it daunting, if not almost impossible, to withstand the humanitarian and environmental crises caused by climate change. The extraction-driven industries built on the platform of colonialism by the so-called "richer" nations of today have been primarily responsible for climate change. Yet these nations have made little attempt to take responsibility and atone for their destructive actions. The reckless capitalist pursuit of growth, production, and profit have propelled some to protect their lavish lifestyles with no regard for the negative consequences of their actions for the poor and vulnerable. The vast disparities in access to resources underpin the structured inequalities we encounter around the world. These inequalities traverse constituencies—all of which are shaped by the other constituencies—that include, but are not limited to, pensioners and senior citizens, urban slum dwellers,

Indigenous peoples, ethnic minorities, women, the young, and rural communities. Instances of climate refugees, or women affected by drought and food shortage that increase their burdens, or Inuit populations displaced by eroding shorelines and melting permafrost are already all too frequent. In addition, millions face the risk of being drowned or losing their freshwater resources as sea levels rise, posing the devastating prospects of loss of both homeland and livelihood for millions of people.

We know that responses to the challenge of climate change *can* take the form of social movements and critical writings that straddle the natural sciences and the human sciences; that move fluently between academia and many other groupings; and that, therefore, make good on the realization that progress in any one aspect of the climate crisis can prompt change in the others. It is perhaps a little too obvious to say that climate justice means social justice. We only have to pause to remember the refugees desperate for water in the refugee camps of the early 21st century, and the droughts that pervade places as disparate as Ethiopia, India, Kenya, Syria, and the US, among others. Of course, all have very differing impacts, depending on people's access to resources for adaptation and recovery, and yet also share a common cause: the lack of climate justice and the lack of social/political/economic justice. Our volume aims to forge connections among these locations, alongside the emerging networks of social movements, NGOs, websites, think tanks, community organizations, and impromptu local struggles that attempt to tackle this "super wicked" problem (Levin, Cashore, Bernstein, and Auld 2012). It is evident that a global climate justice movement continues to grow— witness the million marchers around the globe on November 29, 2015 on the eve of the COP 21 in Paris, France, in a week when the French "State of Emergency" made illegal any political public gatherings of two or more people,[1] and the world-wide school students' strike for the planet in March 2019.

Today's climate crisis has deep historical roots. When the present geological era is named the "Anthropocene," we implicate the entire trajectory of humanity's productive metabolism as the agent of that crisis. Indeed, the very term "Anthropocene" is "a recognition that humans have created the prospect for a systems failure of the planet," which includes climate change, species extinction, freshwater degradation, and industrialized monoculture agriculture (Kurian 2017: 104). Only movements that firmly identify themselves with historical movements for social justice and change, and that arise from this planetary system failure, will be able to mobilize society-wide transformations around the globe. These movements have often based their strategies and tactics on challenging the forces that continue to deepen this historical ecological rift. For example, the iconic Narmada Bachao Andolan (NBA or Save the Narmada Movement), launched in 1985 to challenge the building of mega-dams such as the Sardar Sarovar Dam on the river Narmada in central India, and their devastating human and environmental consequences, succeeded in questioning the World Bank's apparently unthinking support for large-scale projects, and led to fundamental changes in the recognition of the rights of those displaced by development projects in India. The Sardar Sarovar Dam was inaugurated in 2017 despite the movement against it. Yet the NBA continues to work for social justice among marginalized communities (Uniyal 2017), and its vision of a transformative model of development that links social justice, environmental sustainability, and human rights remains as urgent as ever.

The rise of global environmentalism since the mid-20th century is embodied in a range of environmental movements, including cross-national global networks and national and local movements. Many of these movements have formed strategic alliances to halt, for example, large-scale dam building, by challenging destructive resource extraction and expropriation by corporations and nation states. More recently, mobilizing around climate justice to keep fossil fuels in the ground, such as the *Break Free* civil disobedience action across six continents in May 2016, has also sparked a transition to alternative and just energy systems everywhere. Unifying global discourses of "sustainable development" and "planet earth"—however contested or coopted—are marked by the presence of movements grounded in local realities, cultures, and practices.

Climate justice is the articulation of these movements: the term and the social forces that carry it are gaining momentum. Fighting at the front lines, promoting intersectionality, developing coalitions, and infusing a systemic analysis of the roots of the crisis are fundamental to climate justice movements. And so are unapologetic pressures for radical change, based as they are on expressions of commitment, love, and hope. Yet each movement speaks with a distinct voice and works with specific historical missions. Our volume is an expression of all of this. In centering climate justice for all species, indeed, for the planet itself, we open up debates that could allow for strong participation of readers in creating deeply democratic futures.

The volume brings together contributions that offer innovative analyses, engage with future-oriented strategies, and envision new ways of responding to the climate crisis. Together, these may help enhance the resilience and resistance of a range of marginalized and disadvantaged communities. We argue that one way of thinking about the present moment, and the foreseeable future, is to examine democracy and economics within the context of cultural and ideological change. We do this by showcasing writings from public intellectuals and scholars together with grassroots climate activists who are all in search of fresh ideas for realizing climate justice and just climate futures in collaborative ways. In so doing, we seek to spur public interventions, while deepening the understandings of the broad cross-national networks needed for effective action on climate change. By paying special attention to climate justice struggles around the world, the volume foregrounds compelling new narratives and discursive strategies toward our climate futures. We know that it is a climate justice lens that can catalyze public deliberations and actions, while simultaneously creating a synergistic momentum for effective action on an Earth in crisis.

The book is structured into four parts. Part I, "Climate Change, Colonialism, and Capitalism," opens with a call to challenge colonial histories and capitalist/industrialized presents and futures to ensure climate justice, especially for marginalized and Indigenous peoples. This section delineates particular theoretical foundations of the politics and economics of climate change, alongside an analysis of systemic alternatives to current—and dominant—global systems. It also provides a critique of capitalism and its links to the climate crisis; explores the social dimensions of climate change; and analyses the current failures of the political system to respond to issues of justice, recognition, reparation, and redistribution.

Part II, "Climate Change through Lenses of Diversity," approaches the issue of climate justice from a range of diverse perspectives including those of culture, gender, indigeneity, race, and sexuality. In offering feminist and queer perspectives, two chapters highlight the significance of challenging the masculinist ideologies and exclusionary practices that underpin climate-destroying practices and policies that are present all over the planet. Other chapters develop a culture-centered approach to reframe the narratives around climate change adaptation in the contexts of Aotearoa New Zealand and of indigeneity in Sweden, and the need to get past the us-and-them divides and acknowledge how lives of creatures far and near, human and non-human, are interconnected.

Part III, "Social Sciences, Humanities, and Climate Justice" features analyses of climate change and climate justice that draw on insights from sociology, political science, the humanities, and media studies. We consider the humanities and humanistic social sciences (for some, they are united as the human sciences) that can force open a number of visionary windows on climate change. Encapsulated within the human sciences, are environmental and climate justice perspectives, critical development studies, eco-socialism, and community studies, among others. The section ends with an analysis of the potential of deliberative environmental democracy as a necessary aspect of effective climate governance that can address the demands for climate justice.

Part IV, "The Quest for Climate Justice across the World" offers cases and voices from a wide range of climate-impacted regions including from the continents of Africa, North America, Latin America, Oceania, Europe, and Asia. At the many sites and faces of climate vulnerability, from the small island states of the Pacific and Indian Oceans to the eroding shores of Bangladesh, this section covers issues that are emerging in some frontline states.

The volume ends with Part V, "Thinking beyond the Here and Now: Envisioning Many Futures." It proposes approaches to climate activism, local sustainability initiatives, and transformative action plans that respond to the challenges of climate justice and that demand visions of a different future. It is a vision of the future demanded by the School Strike for Climate movement that saw swathes of school children cover city squares and roads in over a hundred countries on March 15, 2019 to call for immediate action on climate change. An open letter to *The Guardian* by young activists led by the inspirational Swedish school student, Greta Thunberg, captured the sentiment very well:

> Young people make up more than half of the global population. Our generation grew up with the climate crisis and we will have to deal with it for the rest of our lives. Despite the fact, most of us are not included in the local and global decision-making process. We are the voiceless future of humanity. We will no longer accept this injustice. We demand justice for all past, current, and future victims of the climate crisis, and so we are rising up. ('Climate Crisis and a Betrayed Generation' 2019)

Our volume expresses its solidarity with the millions of children who will face the brunt of a climate-changed world, and the many who will rise to do something about it. As the final chapter (Porritt, this volume) says, committing to *"a just transition"* to the future "is now *our* last and best hope."

Note

1. France declared a national state of emergency in November 2015 following a series of terrorist attacks in Paris that killed 130 people.

PART ONE

CLIMATE CHANGE, COLONIALISM, AND CAPITALISM

2 | WAY BEYOND THE LIFEBOAT: AN INDIGENOUS ALLEGORY OF CLIMATE JUSTICE

Kyle Powys Whyte

Inuit culture is based on the ice, the snow and the cold ... Therefore when the climate changes and/or warms ... Then our right to culture, our right to educate our children on the land, our right to safety, our right to health all become impacted by these rapid changes. In essence our Right to exist as Inuit as we know it is impacted ... We are a very adaptable people and yet others tend to think that it is our inability to adapt to the modern world that we are facing these challenges of social and health related issues. Not true ... It is the speed and intensity in which change has occurred and continues to occur that is a big factor why we are having trouble with adapting to certain situations. Climate change is yet another rapid assault on our way of life. It cannot be separated from the first waves of changes and assaults at the very core of the human spirit that has come our way. (Sheila Watt-Cloutier, interviewed by the *Ottawa Citizen*, in Robb 2015)

The Treaty Belt is made of two rows of purple wampum beads, and these two rows have the spirit of the Haudenosaunee and the Dutch ... the two purple rows depict two vessels travelling down a river. One, a birch-bark canoe, is for the Haudenosaunee and contains our laws, customs, and way of life. The other, a ship, is for the Dutch and contains their laws, customs, and way of life. The purpose of the Treaty is to recognize that each People is to travel down this river together, side-by-side but each in their own vessel ... The treaty recognizes that the Haudenosaunee and Dutch share the same river, the river of life. We are to help each other, from time to time, as we travel this river together. We are to take care of this river as all of our survival depends on a healthy river. (Brief description of the Haudenosaunee Kaswentha, in Ransom 1999: 27–9)

The insidiousness of climate injustice

In the first epigraph, Watt-Cloutier says that climate-related risks to her people's health, cultural integrity, and economic vitality are intensified through colonial and capitalist domination—"the first waves of changes and assaults." Years of Indigenous testimonies and, more recently, evidence in major scientific reports, bear witness to the relevance of her claim for many Indigenous peoples (Whyte 2017; Jantarasami et al. 2018). The Nisqually Indian Tribe and Quinault Nation, living within the Pacific Northwest region of what is currently the US, depend on salmon and shellfish for cultural, religious, economic, and nutritional health purposes. Yet important habitats are becoming further degraded from climatic and non-climatic factors, including warming waters, ocean acidification, and the ramped up shoreline development of US settler populations. Late Nisqually leader Billy Frank Jr. states that "As the salmon disappear, our tribal cultures, communities and economies are threatened as never before" (Treaty Indian Tribes in Western Washington 2011: 6).

Writing from a Potawatomi, North American perspective, I see Indigenous peoples as often perceiving the burdens of climate-related risks through their experiences of already having been deeply harmed by the economic, industrial, and military drivers behind anthropogenic climate change (Callison 2014; Wildcat 2009; Houser et al. 2001). Historically, US settlers widely displaced, terrorized, and polluted Indigenous communities for the sake of profiting from oil and coal development (Grinde and Johansen 1995; Small 1994; Weaver 1996). In 20th-century Oklahoma, for example, oil development and settler greed polluted the Sac & Fox Tribe's drinking water and energized the systematic effort—dubbed "the Reign of Terror"—to murder with impunity scores of Osage persons (Grann 2017; Royster 1997). To the north in Oklahoma, Gail Small writes, "Like many Cheyenne, I feel as if I have already lived a lifetime fighting [coal] strip-mining. We live with fear, anger, and urgency. And we long for a better life for our tribe" (Small 1994). Haunting similarities exist when we make global comparisons across Indigenous peoples, such as the pollution and violence endured by the Ogoni people from the multi-national oil industry and the nation of Nigeria (Saro-Wiwa 1992).

Fossil fuel industries remain major concerns in recent times. In 2009, 50 people were killed, many Indigenous, in a conflict over the Peruvian government's endorsement of mining and oil drilling in the Amazon (Aquino 2009). Oil and gas pipelines in North America traverse

Indigenous territories without genuine Indigenous consent to the construction or continued operations of these pipelines, including the now notorious construction of the Dakota Access pipeline in Standing Rock Sioux territory (Dhillon and Estes 2016). Even for Tribes pressured by the US into reliance on fossil fuel industries, such as the Crow Tribe of Indians, the economic dependence has yet to make their members well-off by US standards (Beeler 2017; Turkewitz 2017).

In my experiences, most Indigenous peoples have complicated stories to tell about anthropogenic climate change that often start with their being harmed by fossil fuel industries. The stories continue on to discuss how current laws and policies render them more vulnerable to climate change impacts. The Indigenous peoples of the Isle de Jean Charles in coastal Louisiana, including people who identify as members of the Isle de Jean Charles Biloxi-Chitimacha-Choctaw Tribe and the United Houma Nation, were forced onto a highly vulnerable small island to make way for climate driving industries, including petroleum and industrial agriculture. The US and these industries dramatically engineered the lands and hydrology of the region, which also worked to make the island itself less habitable. The island ecosystem's capacity to support Indigenous cultural integrity, economic development and physical and mental health is curtailed due to pollution and loss of wetlands and barrier islands that protect against extreme weather. Now climate-related sea level rise is a major concern. In terms of law and policy, the US still fails to recognize the Indigenous peoples there as politically self-determining sovereigns. Federal and local agencies and contractors working on regional disaster response planning have not achieved adequate Indigenous representation or free, prior, and informed consent in the decision-making and planning processes (Dardar 2012; Maldonado et al. 2013; Crepelle 2018). Former Principal Chief Thomas Dardar Jr. testified to US Congress that "Unfortunately the United Houma Nation's pursuit of federal recognition is closely tied to the repetitive disasters we faced" (Dardar 2012). For the Indigenous peoples of the island, the United Nations (UN) *Declaration on the Rights of Indigenous Peoples* (2008) expresses rights to enjoy the types of goods threatened by climate change and extractive industries, including political self-determination, cultural integrity, economic development, and high standards of physical and mental health.

Even strategies for lowering national carbon footprints pose risks to Indigenous peoples and put their human rights in peril, whether through programs of the World Bank, the United Nations, or particular nations.

Hydropower and forest conservation *still* involve displacement of Indigenous peoples (Beymer-Farris and Bassett 2012; Cooke et al. 2017). Wide-ranging technological solutions, from natural gas transitions to permanent nuclear waste storage to wind power to geoengineering, pose significant risks that include desecration of sacred sites, pollution, unequal economic profits, violations of free, prior, and informed consent, and sexual violence through sex trafficking at *man camps* set up to house oil and gas industry workers (Whyte 2018; Deer and Nagle 2017; Dussais 2014; Eaton 2017; Endres 2009; Bronin 2012). A report by Tauli-Corpuz and Lynge (2008) for the UN Permanent Forum on Indigenous Issues documents numerous rights violations and risks involved with climate change mitigation measures, from biofuels to forest conservation to hydropower.

Public discourses of Indigenous allies, including climate scientists and journalists, can also be problematic when they portray Indigenous vulnerability to climate change without reference to the larger struggles with colonialism and capitalism I have described so far (Cameron 2012). Such discourses give the false impression that Indigenous peoples are isolated populations that face risks only because climate change, via bad luck, happens to affect the flora and fauna they depend on. Ignoring colonial and capitalist domination also impacts negatively the attempts of some climate scientists to collaborate with Indigenous peoples to learn about climate change. For Indigenous peoples' knowledge of climate baselines or shifting species ranges can divulge sensitive information about sacred, medicinal, or economically valuable species and ecosystems that are still threatened by the actions of settlers, corporations, and nations or are not sufficiently protected under current legal frameworks (Williams and Hardison 2013). Research organizations, with legacies of exploitative research against Indigenous peoples, often still do not create career incentives and educational opportunities to enable scientists to work with their Indigenous collaborators in ways that reduce risks and create mutual benefits.

So far, I have covered a complex landscape of Indigenous climate justice involving colonial and capitalist domination linked to industrialization. Is there a succinct way to convey an Indigenous perspective on climate justice that makes all these connections? Indigenous persons sometimes turn to allegory to illustrate justice, such as Dean Suagee's (1994) *Turtle's War Party: An Indian Allegory of Environmental Justice.* Honig (1994) too provides an account challenging the capacity of some moralizing stories to address the complexity of injustice and moral decision-making. Perhaps an allegorical story of vessels can be created to describe Indigenous climate justice.

Elizabeth Deloughery's (2007) research shows how narratives and metaphors of ocean voyaging in the Pacific are created to describe both cases of oppressive and liberatory relationships across different persons and societies who share land, water, and air. Christina Sharpe (2016) weaves images, expressions and histories of ships and their wakes, such as the Middle Passage, to illustrate complex aspects of anti-Black racism, where *wake* involves the interplay of multiple meanings, including tracks on water, currents of air, mourning rituals, and consciousness (*a*wake). Buckminster Fuller's "spaceship earth" flies through space without the possibility of getting more fuel and supplies, illustrating dependence on finite resources (Fuller 2008). Vicente Diaz (2011) invokes Austronesian canoes, archipelagos, and seafaring to envision fluid practices that liberate Indigenous peoples from the containment and isolation of colonialism. Garrett Hardin describes rich countries as lifeboats surrounded by poor people swimming in the surrounding oceans—there is only so much room on the lifeboat for poor people as environmental and economic conditions deteriorate (Hardin 1974). Martin Luther King, Jr., in an attempt to motivate respect for diversity and justice in the US, said the widely quoted phrase "We may have come to these shores on different ships, but we are now all in the same boat."

The second epigraph at the beginning of this chapter features an enduring story of vessels, the Treaty Belt, or *Kaswentha*, between the Haudenosaunee and Dutch from the 1600s, that illustrates their relationship through depicting the politics of sharing a river. To illustrate the relationship between colonialism, capitalism, industrialization, and climate injustice, I have written an allegory of vessels, inspired by the *Kaswentha*, but very different in its telling.

An Indigenous allegory of climate justice

Imagine a world of many vessels floating in a pool of waters under a sky. Humans, animals, and plants live on these vessels; some, such as fish, live in the waters too. The vessels are all built very differently based on the histories, geographies, economic statuses, cultures, and aspirations for the future of the occupants. Some vessels are birch-bark or other canoes; others are ships or sail boats. Each vessel is its own amalgamation of many different episodes of collective human and plant/animal lives and histories.

Some vessels will nonetheless be referred to as canoes given it is possible to see the traces of that boat-making style in their construction. To be accurate though, they are more like conglomerations of canoes with smaller boats such as speed boats and zodiacs, each connecting to one another, like a very complicated catamaran with many moving pontoons. Other vessels are large aircraft carriers with massive towers built on top of them; they also have upside-down looking towers built on their ship bottoms that extend deep underwater nearing the pool's floor. Many of the vessels are intricately connected to each other in various relations of interdependence. There are millions of different power lines, bridges, ropes, shuttles, and other materials that connect the vessels to each other. Now imagine what this looks like as myriad people and nonhumans regularly move from one vessel to the next.

Aircraft carriers have towers and high-technology equipment on them. Yet aircraft carriers are the descendants of large wooden balloon carriers that were built using the wooden materials of canoes that were destroyed by the ancestors of some of the current aircraft carrier occupants. Glorious paintings of these balloon carriers hang in the aircraft carrier rooms and some of the old furniture in these rooms still has refurbished canoe wood. The watery depths and pool floor below where the early canoes were destroyed furnished many of the resources, such as metals and fossil fuels, which were used to transition from the balloon carriers to the current aircraft carriers. The canoes that avoided being absorbed by aircraft carriers and their ancestor vessels still float in the water. But they are much lower in height than the aircraft carriers. Occupants of the canoes can see what is going on in the water in great detail and observe firsthand trends in water quality and turbulence.

Residents of the aircraft carriers, on the other hand, rarely get close to the water. They observe the water through windows, probes, and submarines. Canoes and aircraft carriers are both dependent on the water conditions for the well-being of the people inside them, but the perspectives of the occupants differ based on whether they can experience the waters with an immediacy or whether they must rely on experiencing the waters based on reports from probes sent out by the aircraft carriers. Many of the aircraft carriers have forcibly attached themselves to the canoes. Some people who were born on the canoes now live on the aircraft carriers, bringing with them shards of materials from the canoes that they often have to sell for food. However—and with some exceptions—these persons live on the parts of the aircraft carriers that are most exposed to the water or are more likely to be flooded.

The aircraft carriers often seek to get rid of the canoes when the canoes are in the way of a new addition to the aircraft carrier. They do so by pulling in the canoes, smashing them up against the sides of the aircraft carrier, absorbing any surviving occupants and taking any valuable supplies or shards from the canoes. Sometimes, occupants of aircraft carriers are curious and seek to experience the waters from the vantage points of the canoes. Yet, there is always a risk that too many inquisitive occupants of aircraft carriers can overwhelm a canoe's carrying capacity and supplies, either when the aircraft carrier occupants all try to board the canoe at the same time or when even just a few leap dramatically from an aircraft carrier to a canoe.

The strangest vessels of all are neither canoes nor aircraft carriers. They are not so much water-going vessels as they are giant hovercrafts that float above all the other vessels in the sky. The force of the fans of the hover-engines blows into the waters. Canoes, aircraft carriers, humans, plants, and animals bob underneath the hovercrafts like billions of marionettes as the hovercrafts seek to take their resources from the waters as well as from the floor of the pool. The hovercrafts are massive in scale, spanning across many vessels, often blocking the sun in the sky just as an eclipse does, especially when the hovercrafts get close to each other. The hovercrafts often tie up the canoes like yoyos, smashing them up against the aircraft carriers. Sometimes, they drop stone weights onto the canoes, weights that plunge the canoes into the bottom of the waters. In the turbulence that is created, canoes take in water that needs to be bailed out. The hovercrafts often attach on to the aircraft carriers to prop them up when the hulls of the aircraft carriers get punctured.

While water is a source of sustenance for all people, it can also pose a threat to the cultural, economic, and political self-determination of the people in the vessels, especially if the vessels take in water, or sink as a result of turbulence. The aircraft carriers create turbulence in the waters through the wake of their engines, and their sheer weight in the water. The turbulence is made worse by the blowing of the giant fans of the hovercrafts. The canoes, depending on their location and connections to aircraft carriers or hovercrafts jostle and bristle more in the water than the other vessels. Without the aircraft carriers and hovercrafts, the water would be a lot calmer and canoes would have a lot more control, and be more stable.

The people in the aircraft carriers and hovercrafts no longer see how much turbulence their vessels create for canoes, because few of them live or work close enough to the water. The canoes simply disappear from vision or appear as small, fragile boats that are so vulnerable to the waters that it feels as if nobody would dare go near them to find out what is happening. Nor do many of the people in the aircraft carriers see the risks endured by their fellow residents who live in areas of the ship that are more exposed to the waters. Moreover, going back many years, the aircraft carriers and hovercrafts have destroyed the aquatic carbon sinks and used fossil fuels so that they could power their engines and fans, producing greater turbulence in the waters, and more intense and frequent storms. The smoke stacks of the hovercrafts are piled on top of the crafts, with a direct line to polluting the sky. Underground oil drilling and leaking pipelines make the turbulent waters dirty as well.

Given that many occupants in the canoes can actually see the water, because the canoes are lower in height, those occupants have a good sense of how bad the turbulence is, how dirty the water is, and what to expect from the storms. People living high up inside aircraft carriers or hovercrafts have not ever been close to the water, yet the technologies at their disposal allow them to reach out into the water and test for trends of greater or lesser turbulence. As the turbulence grows and the storms become more intense, all vessels are affected, but the canoes bear the brunt. Some sink completely into the water, their occupants escaping onto other canoes or, at times, onto aircraft carriers; others change their location in the water and detach from the aircraft carriers, in the process facing the onslaught of the disturbance caused by the hovercraft engines.

Some of those on the aircraft carriers realize that they are responsible for stirring up the waters or at least realize that they should do something about the impacts of the turbulence on smaller vessels. The occupants of the hovercrafts realize this, too; however, they feel that the solution is for them to create more, or larger, hovercrafts, eventually moving everyone into a network of interconnected hovercrafts or on to a massive, single hovercraft. The aircraft carriers, often disagreeing with the somewhat impractical ideas of the hovercrafts, put up proposals to bring the canoes closer to each other—in some cases lifting them out of the water and hoisting them like lifeboats; in other cases, beams and other materials are added to stabilize the canoes in the water. Yet, in cases where the aircraft carriers' proposals are put into practice, the results are disastrous. When the aircraft carriers fire their engines to move closer to the canoes, the wake of their engines further destabilizes the water, harming the canoe occupants even more. When severe storms break, the aircraft carriers cannot even see the canoes and run right over them.

The aircraft carriers blame climate change and smallness of the canoes for this unfortunate state of affairs. Yet, the occupants of canoes have a different perspective. The reason why the aircraft carriers cannot get closer is because they are too big, and their engines disturb the water greatly. The solutions with aircraft carriers and hovercrafts always involve creating more turbulence, a turbulence that was there even before "climate change" as an issue stirred the waters. One way to lessen the turbulence and storminess is to change the design of the aircraft carriers and hovercrafts completely. Yet this would mean that the occupants of those vessels would have to redesign their vessels in ways that do not disturb the waters. If there are parts and mechanisms of the aircraft carriers and hovercrafts that are not needed any more because they always disturb the waters, what would the occupants do with those parts and mechanisms? What would the occupants do in their lives without those parts and mechanisms?

Interpreting the allegory

There are certainly many ways to interpret the allegory. The following is simply one place to begin—a starting point I offer as I quickly close this piece. The canoes represent the many different Indigenous peoples everywhere and people who share their situation. The aircraft carriers are nation-states and the hovercrafts are corporations. The pool of waters and sky are the earth system at broad and local scales. The engines, fans, and carbon-intensive economics of the aircraft carriers and hovercrafts, their sheer size and desire to clear out the canoes, represent the nexus of colonialism, capitalism, and industrialization. This nexus has multiple forms, including the military, extractive industries, and educational institutions. Capitalism and industrialization stir and disrupt the water and sky; colonialism makes it hard for some vessels to adapt to the disruptions.

I offer the allegory to demonstrate why climate change and many proposed solutions to adapt to or mitigate climate change produce great suffering for Indigenous peoples *unless* colonialism is addressed alongside capitalism and industrialization. Failing to address colonialism is like lowering or turning off some of the engines and fans of the aircraft carriers and hovercrafts. This, in turn, would sink the large vessels, pulling down the canoes with them. Given the physical realities of the vessels and waters in the allegory, it is hard to see how tweaks in the ways things work would change the ultimate ecological dynamics that are so unfavorable to the canoes.

So Watt-Cloutier's references to capitalism and colonialism in the opening quotations should not be dismissed within movements for climate justice. In the absence of a concern for addressing colonialism, climate justice advocates do not really propose solutions to climate change that are that much better for Indigenous well-being than the proposed inaction of even the most strident climate change deniers. Decolonization and anti-colonialism, understood in senses appropriate to the allegory, cannot be disaggregated from climate justice for Indigenous peoples. Indigenous climate justice movements are distinct in their putting resistance to the nexus of colonialism, capitalism, and industrialization at the vanguard of their work.

3 | THE POLITICS OF CLIMATE CHANGE IS MORE THAN THE POLITICS OF CAPITALISM[1]

Dipesh Chakrabarty

Many of us still approach the problem of global warming armed only with weapons forged in times when globalization (of media, capital) seemed to be the key issue for the world. Globalization and global warming are no doubt connected phenomena, capitalism itself being central to both. But they are not identical problems. The questions they raise are often related, but the methods by which we define them as problems are, equally often, substantially different. Social scientists, especially friends on the left, sometimes write as though these methodological differences did not matter; that scientists are, after all, only studying or measuring the outcomes of capitalism while we, with our methods of political economy, always knew what the ultimate cause of it all was! What I wish to do in this brief statement is go over some of the narratives that the findings of natural or biological sciences make possible. It is not my aim in this short essay to resolve the tensions I point to in our narratives of climate change.

Two approaches to climate change

One generally finds two approaches to the problem of climate change. One dominant approach is to look on the phenomenon simply as a one-dimensional challenge: How do humans achieve a reduction in their emissions of greenhouse gases (GHGs) in the coming few decades? The question is driven by the idea of a global "carbon budget" that the fifth aggregate report of the Intergovernmental Panel on Climate Change (IPCC) foregrounded. It also sets as its target the idea of keeping the average rise in the surface temperature of the planet *below* the 2°C threshold, since anything above that is labeled "dangerous." The climate problem is seen in this approach as a challenge of how to source the energy needed for the human pursuit of some universally accepted ends of economic development, so that billions of humans are pulled out of poverty. The main solution proposed here is for humanity to make a transition to renewable energy as quickly as technology and market signals permit. The accompanying issues of justice concern relations between poor and rich nations and between present and future generations. For example, should not the less developed and more populous countries (like China and India) have a greater right to pollute, while the developed nations take on more responsibility to make deep cuts in their emissions and undertake financial commitments to help the developing nations achieve their goals? The question of how much sacrifice the living should make as they curb emissions, to ensure that unborn humans inherit a world that enables them to achieve a better quality of life than the present generation, remains a more intractable question, and its political force is reduced by the fact that the unborn are not here to argue about their share of the atmospheric commons.

Most imagine the problem to be mainly one of replacing fossil fuel-based energy sources by renewables; many also assume that the same modes of production and consumption of goods will continue. These latter analysts imagine a future in which the world is more technologically advanced and connected than now, but with the critical difference that a consumerist paradise will be within the reach of most, if not all, humans. Some others—on the left—would agree that a turn to renewables is in order, but argue that since it is capitalism's constant urge to "accumulate" that has precipitated the climate crisis, the crisis itself provides yet another opportunity to renew and reinvigorate Marx's critique of capital. There is clearly an assumption that a globalized, crowded (9–10 billion people), socially just, and technologically connected post-capitalist world can somehow come into being and avoid the pitfalls of the drive to accumulate. And then there are those who think of not just transitioning to renewable sources of energy but of actually scaling back the world economy, de-growing it, and thus reducing the ecological footprint of humans while desiring a world marked by equality and social justice for all. Still others think—in a scenario called "the convergence scenario"—of reaching a state of economic equilibrium globally whereby all humans live at more or less the same standard of living. And then, of course, there are those who think of the most desirable future as capitalist or market-based growth *with* sustainability.

Against all this, there is another way to view climate change: as part of a complex family of interconnected problems, all adding up to the larger issue of a growing human footprint on the planet that has, over the last couple of centuries, seen a definite ecological overshoot on the part of humanity. This overshoot, of course, has a long history but one that has picked up pace in more recent times. The Israeli historian Yuval Noah Harari explains the issue well. It is only "in the last 100,000 years," says Harari, "that man jumped to the top of the food chain" (2015: 9). This has not been an evolutionary change. As Harari explains:

> Other animals at the top of the pyramid, such as lions and sharks, evolved into that position very gradually, over millions of years. This enables the ecosystem to develop checks and balances that prevent lions and sharks from wreaking too much havoc. As the lions became deadlier, so gazelles evolved to run faster, hyenas to cooperate better, and rhinoceroses to be more bad-tempered. In contrast, humankind ascended to the top so quickly that the eco-system was not given time to adjust. (2015: 11–12)

The problem of humans' ecological footprint, we can say, was ratcheted up with the invention of agriculture (more than 10,000 years ago) and then again after the oceans found their present level about 6,000 years ago and we developed our ancient cities, empires, and urban orders. It was ratcheted up yet again over the last 500 years with European expansion and colonization of faraway lands, and the subsequent rise of industrial civilization. But a further ratcheting up by several significant notches happened after the end of the Second World War, when human numbers and consumption rose exponentially thanks to the widespread use of fossil fuels, not only in the transport sector but also in agriculture and medicine allowing, eventually, even the poor of the world to live longer—though not healthy—lives.

GHG emissions gave humans the capacity to interfere in Earth Systems, yielding the planetary-scale geological agency that quite a few scientists and science-scholars including David Archer (2009: 6) and Naomi Oreskes (2007: 93) have written about. This planet-wide geological agency of humans, however, cannot be separated from the way humans interfere in the distribution of natural life on the planet. Not only have marine creatures not had the evolutionary time needed to adjust to our newfound capacity to hunt them out of existence through deep-sea fishing technology, but our GHG emissions now also acidify the oceans, threatening the biodiversity of the great seas, and thus endangering the very same food chain that feeds us. Jan Zalasiewicz (2015) and his colleagues on the sub-committee of the

International Stratigraphy Commission, charged with documenting the Anthropocene, are thus absolutely right to point out that it is the human record left in the rocks of this planet as fossils that will constitute the long-term record of the Anthropocene, perhaps more so than the excess GHGs in the atmosphere.

Viewed thus, the idea of the Anthropocene increasingly becomes more about the expanding ecological footprint of humanity as a whole—and this must include the question of human population, for while the poor do not have a direct carbon footprint, they contribute to the human footprint in other ways (this is not a moral indictment of them)—and less about a narrowly defined problem of climate change.

> The climate change problem is not a problem to be studied in isolation from the general complex of ecological problems that humans now face on various scales—from the local to the planetary … There is no single silver bullet that solves all the problems at once … What we face does indeed look like a wicked problem, one that we may diagnose but not be able to "solve" once and for all. (Chakrabarty 2016b: 106–7)[2]

The Anthropocene and the inequities of capitalism

In my essay *The Climate of History: Four Theses*, I acknowledged that there was "no denying that climate change has profoundly to do with the history of capitalism" but added that it could not be reduced to the latter (Chakrabarty 2009: 212). I then went on to point out that while climate change would only accentuate the inequities of the global capitalist order as the impact of climate change—*for now and in the immediate future*—falls more heavily on poorer nations and on the poor of the rich nations, it was different from the usual crises of capital. I said: "Unlike in the crises of capitalism, there are no lifeboats here for the rich and the privileged" (Chakrabarty 2009: 221).

Many scholars on the left vehemently oppose the idea this could be a crisis for all of humanity; hence they criticize the expression "human"-induced climate change. Thus, Swedish academics Andreas Malm and Alf Hornborg (2014) ask in a widely cited essay that if human actions have indeed precipitated this collective slide into a geological period that signifies human domination of the planet and even of its geological history, then why name that period after all humans or the human species, the *anthropos*, when we know it is the rich among humans or the institutions of capitalism or the global economy that are causally (hence morally?) responsible for this change in our condition? "A significant chunk of humanity is not party to fossil fuel at all," they point out, and add: "hundreds of millions rely on charcoal, firewood or organic waste such as dung" (2014: 65). They cite the Canadian scholar Vaclav Smil to say that "the difference in modern energy consumption between a subsistence pastoralist in the Sahel and an average Canadian may easily be larger than a 1,000-fold," hence "humanity seems far too slender an abstraction to carry the burden of causality [for climate change] ... Realizing that climate change is 'anthropogenic' is really to appreciate that it is *sociogenic*" (2014: 65). They then go on to criticize my statement regarding the rich having no "lifeboats." "[T]his is a flawed argument," they write. "It blatantly overlooks the realities of differentiated vulnerability on all scales of human society ... For a foreseeable future—indeed, as long as there are human societies on Earth—there *will* be lifeboats for the rich and the privileged" (2014: 66). Quite a few other scholars have since repeated the charge.

I find it ironic that some scholars on the left should speak with an assumption similar to that made by many of the rich, who do not necessarily deny climate change but believe that, *whatever the extent of the warming* and destabilization of the climate, they will always be able to buy their way out of the problem! This is understandable coming from economics textbooks that envision capitalism as an economic system that will always face periodic crises and overcome them, but never face a crisis of such proportions that it could upset all capitalist calculations. It is easy to think within that logic that climate change was just another of those business-cycle type challenges that the rich had to ride out from time to time. Why would scholars on the left write from the same assumptions? Climate change is not a standard business-cycle crisis. Nor is it a standard "environmental crisis" amenable to the usual risk-management strategies.

Left unmitigated, climate change affects us all, rich and poor. They are not affected in the same way, but they are all affected. A runaway global warming leading to a Great Extinction event will not serve the rich very well. A massive collapse of human population caused by climate dislocation—were it to happen—would no doubt hurt the poor much more than the rich. But would it not also rob global capitalism of its reserve army of "cheap" labor on which it has so far depended? A world with freakish weather, more storms, floods, droughts, and frequent extreme weather events cannot be beneficial to the rich who live today or to their descendants who will have to live on a much more unfriendly planet. In fact, the journal *Science News* has just reported the conclusions of a study led by Professor Sergei Petrovskii of their department of applied mathematics that suggests that "an increase in the water temperature of the world's oceans of around six degrees Celsius—which some scientists predict could occur as soon as 2100—could stop oxygen production by phytoplankton by disrupting the process of photosynthesis" (University of Leicester 2015: n.p.; see also Sekerci and Petrovskii 2015). Not a great prospect, even for the super-rich.

Of course, this is an extreme scenario. But the point of the lifeboat metaphor was not to deny that the rich, depending on how rich they are, will always have—compared to the poor—more resources at their disposal to deal with disasters and buy their way to relative safety. It is possible that the lifeboat metaphor was too cryptic (and it clearly misfired for some readers) but my point was that climate change, potentially, has to do with changes in the boundary conditions needed for the sustenance of human and many other forms of life. The rich, for all their money, for example, would not find it easy to live in a world whose supply of oxygen had dried up; even they are subject to biological processes! And, to stay with the polemics for a moment, it could be argued that even the super-rich need functioning markets and technological systems to continue to enjoy the benefits of their wealth and investments. In the extreme—and let us hope, unlikely—scenario of runaway global warming, the descendants of the super-rich will find it difficult to hold on to their privileges.

Consider also this additional argument: if the rich could simply buy their way out of this crisis and only the poor suffered, why would the rich of the rich nations do anything about global warming unless the poor of the world (including the poor of the rich nations) were powerful enough to force them to act? Such power on the part of the poor is clearly not in evidence. Nor were the rich nations ever known for their altruism. A better case for rich nations and classes to act on climate change, it seems to me, is couched in terms of their enlightened self-interest. The science of global warming allows us to do so by precisely making the point that, for all its differential impact, it is a crisis for the rich and their descendants as well—as Hansen's (2009) popular book amply makes clear. Besides, some rich nations like Australia lie very exposed to the likely negative impacts of climate change. So yes, a politics of even broader solidarity than simply solidarity of the poor is called for, though I agree that this is by no means easy to achieve.

Politics in/of the Anthropocene

So long as we think of climate change simply as a problem of greenhouse gas emissions and as a matter of transitioning to renewables within a given timetable and a specified carbon budget, we can also point to what might constitute "the politics of climate change," for example the just distribution of the carbon budget between developed and emergent economies and poorer or more immediately threatened nations. A very difficult question to ponder, however, is whether or not the climate crisis—when seen as symptomatic of humanity's ecological overshoot—also signals the first glimpse we might have of a possible limit to our very human-centered thinking about justice, and thus to our political thought as well. Global warming accentuates the planetary tendency towards human-driven extinction of many other species, with some scientists suggesting that the planet may have already entered the beginnings of a long (in human terms) Great Extinction event (Ceballos et al. 2015).

Our political and justice-related thinking remains very human-focused. We still do not know how to think conceptually—politically or in accordance with theories of justice—about justice towards non-human forms of life, not to speak of the inanimate world. Thinkers of animal rights have extended questions of justice towards some animals, but their theories are limited by strict requirements relating to the threshold of sentience in animals. Besides, some philosophers also argue that, whatever the practical value of a category such as life in biology, "life as such" cannot be a strict philosophical category. Yet we cannot think "extinction" without using the category "life," however difficult it may be to define it. The really difficult issue that arises when scholars write about humans being stewards of the planet is what our relationship, conceptually, would be to bacteria and viruses, given that many of them are not friendly to the human form of life (while many are).

So while I agree that politics as we know it continues and will continue into the foreseeable future, and that there is no politics of the Anthropocene as such (but much politics about the label "Anthropocene"!), a deepening of the climate crisis and of the ecological overshoot of which it is a symptom may indeed lead us to rethink the (European) tradition of political thought that has, since the 17th century and thanks to European expansion, become everybody's inheritance today. Nigel Clark makes a similar point from a somewhat different point of view:

A generous—and apposite—response to Anthropocene inquiry, then, might be a new willingness in critical, social, cultural and philosophical thought to embrace the fully *inhuman* … [and] to connect up with … vast domains that are themselves recalcitrant to the purchase of politics. In this way, the Anthropocene … confronts the political with forces and events that have the capacity to undo the political. (2014: 27–8)

Species thinking

Now back to the question of whether or not we should think of humans through the biological category of "species," alongside other historical categories such as "capitalism." Malm and Hornborg take the position that while "the Anthropocene" might effectively represent a possible polar-bear point of view—since they, the bears, might want to know "what species is wreaking such havoc on their habitats," "(w)ithin the human kingdom … species-thinking on climate change is conducive to mystification and political paralysis" (2014: 67). Let me say why I disagree. Can the story of ecological overshoot by humans be thought of simply as the story of modernization and its inherent inequalities and also not as the story of a particular species—*Homo sapiens*—coming to dominate the biosphere to such an extent that its own existence is now challenged? Think of the story as Harari tells it. Today, with their consumption, numbers, technology, and so on, humans yes, all humans, rich and poor—put pressure on the biosphere (the rich and poor do it in different ways and for different reasons). Harari puts the point well:

Humankind ascended to the top [of the food chain] so quickly that the ecosystem was not given time to adjust. Moreover, humans themselves failed to adjust. Most top predators of the planet are majestic creatures. Millions of years of domination have filled them with self-confidence. Sapiens by contrast is more like a banana republic dictator. Having so recently been one of the underdogs of the savannah, we are full of fears and anxieties over our position. (2015: 11–12)

He concludes:

Many historical calamities, from deadly wars to ecological catastrophes, have resulted from this over-hasty jump. (2015: 12)

If one could imagine someone watching the development of life on this planet on an evolutionary scale, they would have a story to tell about *Homo sapiens* rising to the top of the food chain within a very, very short period in that history. The more involved story of rich–poor differences would be a matter of finer resolution in that story. As I have said elsewhere, the ecological overshoot of humanity requires us to both zoom into the details of intra-human injustice—otherwise we do not see the suffering of many humans—and to zoom out of that history, or else we do not see the suffering of other species and, in a manner of speaking, the suffering of the planet (Chakrabarty 2016a: 189–99). Zooming in and zooming out are about shuttling between different scales, perspectives, and different levels of abstraction. One level of abstraction does not cancel out the other or render it invalid. But my point is that the human story can no longer be told from the perspective of the 500 years (at most) of capitalism alone.

Humans remain a species in spite of all our differentiation. Suppose all the radical arguments about the rich always having lifeboats and therefore being able to buy their way out of all calamities including a Great Extinction event are true; and imagine a world in which some very large-scale species extinction has happened and that the survivors among humans are only those who happened to be privileged and belonged to the richer classes. Would not their survival *also* constitute a survival of the species (even if the survivors eventually differentiated themselves into, as seems to be the human wont, dominant and subordinate groups)?

The ecological overshoot of humanity does not make sense without reference to the lives of other species. Nor does it amount to the claim that any one particular discipline now has the best grip on the experience of being human. Biology or something that misses out on the existential dimension of being human will never capture the human experience of falling in love or feeling love for God in the same way that poetry or religion might. A big brain gives us a capacity for cognition of that which is really big in scale. But it also gives us our deeply subjective experience of ourselves and our capacity to experience our individual lives as meaningful. We cannot produce a consilience of knowledge. But surely we can look upon ourselves and on the human story from many perspectives at once.

Debating climate change in uneven public spheres

Climate change is an unfolding problem, and human responses to it—both practical and intellectual—will no doubt vary with the actual futures we come to face. Ten years ago, before the fourth assessment report of the IPCC became the subject of great publicity in print and electronic media, a typical laundry list of debatable questions with reference to climate change would have seemed rather different and much less urgent than issues about the climate that agitate us today. Ten years ago, it was difficult, for example, to interest social scientists in India—the country I am from and a country that is among the top four biggest emitters of greenhouse gases today—in the topic of climate change.

Everyone, however, was absorbed in debating globalization. Foucault and Agamben, governmentality and bio-politics, and the economists Sen, Stiglitz, and Bhagwati, were on everyone's lips, not Paul Crutzen, Eugene Stoermer, or the idea of the Anthropocene.

The first essay I ever wrote on climate change—*The Climate of History: Four Theses*—was written originally in Bangla (Bengali) in a Calcutta journal, *Baromas*, in 2008. No one in the city (or elsewhere) took much notice of it until I translated and expanded it into an English version for the American journal *Critical Inquiry*, which published it in 2009. The experience made me aware of two aspects of the contemporary world I inhabit. Not all global issues were equally global. Globalization—including questions about multinationals, money markets, derivatives and complex financial instruments, the net, the social media, and, of course, the global media—was a genuinely global topic that was discussed everywhere but global warming was not. And it also became clear who set the terms of the discourse. It was the scientists of nations that played a historical role in precipitating the problem of global warming through their emission of polluting greenhouse gases—for example, the United States, the United Kingdom, Australia, and other developed countries—who played two critical roles: as scientists, they discovered and defined the phenomenon of anthropogenic climate change, and as public intellectuals they took care to disseminate their knowledge so that the matter could be debated in public life in an informed manner. I am thinking of scholars/researchers like James Hansen, Wallace Broecker (who coined the phrase "global warming"), Paul Crutzen, Jan Zalasiewicz, David Archer, Will Steffen, Tim Flannery, and others. Scientists of emerging economies like China and India remained confined to their specialist arenas of research. None of them, to my knowledge, wrote any book to explain global warming for the general reader. Global warming is a planetary phenomenon. But as a subject of discussion, it seemed to be distributed very unequally in the world. The situation has changed somewhat in the last ten years—thanks in part to the increasing frequency and fury of extreme weather events in different areas of the world—but not substantially.

What are the implications of this disparity in the distribution of information? It surely skews the "global" debate on climate change in more than one way. When governments come to global forums to discuss and negotiate global agreements on climate change, they do not come equally resourced with informed public discussions in their respective nations, while some governments, admittedly, do not even desire informed publics. More importantly, it means that our debates remain anchored primarily in the experiences, values, and desires of developed nations, that is, in the West (bracketing Japan for the moment), even when we think we are arguing against what we construe to be the selfish interests of "the West."

Notes

1. A version of this chapter first appeared under the same title in *Theory, Culture and Society* 34(2–3): 25–37 in 2017.

2. See the detailed and excellent discussion by Frank P. Incropera (2016).

4 | THE GRAND THEFT OF THE ATMOSPHERE: SKETCHES FOR A THEORY OF CLIMATE INJUSTICE IN THE ANTHROPOCENE

Andreas Malm and Rikard Warlenius

There is a half-minute film clip from the torrential downpours that struck Peru in March 2017, carrying away homes, tearing apart bridges, sending people fleeing from slums in the capital, in an event so unusual in the country as to be explicable only by reference to the exceptionally high temperatures—five to six degrees Celsius above normal—in the nearby Pacific Ocean (Collyns and Watts 2017). The clip shows an enormous river of mud and debris from collapsed houses violently rushing forth over breached banks. Everything is the color of a clayey brown. Suddenly the holder of the camera screams *"hay una persona!"*—there is a person!—and zooms in on some floating planks. The shape of a woman emerges, covered in brown like the rest of the wreck, trying to stand up; she moves with the slowness of a zombie waking up from the dead. Balancing on the planks, falling into the mud, she is finally reached by a group of people on the shore and pulled out of the sludge. Then the clip ends. The number of casualties were around 70, the number of villages destroyed 1,000, the number of homes demolished 100,000, and the number of homeless 1 million.

Indeed, in early 2017, the ravages of the climate imposed states of emergency across the Global South—in Peru, the downpour followed extreme drought and wildfires; in Bangladesh, villages were silently abandoned to the rising sea; in Egypt, the salty ocean pushed ever further into the delta soil, and so on—while Donald Trump instructed American industry to dig up and burn as much coal as possible. It seems the planet has been locked into an infernal furnace. Who put it there and threw away the keys? One common answer is: all of us. Writers, academics, politicians, business leaders are wont to say that we, you and I, even the human species as a whole, are responsible for the leap into this fire. In one of the most high-profile books on the topic published in 2016, *The Great Derangement: Climate Change and the Unthinkable,* celebrated

Indian novelist Amitav Ghosh offers an unusually barefaced version of the answer: global warming "is the unintended consequence of the very existence of human beings as a species." More than that, it is

> the product of the totality of human actions over time. Every human being who has ever lived has played a part in making us the dominant species on this planet, and in this sense every human being, past and present, has contributed to the present cycle of climate change. (Ghosh 2016: 115, see also 32)

On this view, the woman in the mud has her own responsibility for the catastrophe in which she is caught, merely by dint of being *una persona*. Furthermore, that responsibility includes her mother and father and all her ancestors, back into the earliest mist of time. Then there cannot be anything particularly unjust or unfair about her fate, or that of the poor Peruvians whom no one could save: it's their very own chicken coming home to roost, the chicken bred by them and all other members of the biological entity known as *Homo sapiens*. But it is, of course, rather unclear what foundations such a view can have. If we take it seriously, it implies that Australian Aborigines who lived in windbreaks, hunted kangaroos with wooden sticks, and caught possums with rope some 30,000 years ago somehow participated in the large-scale combustion of fossil fuels. So did a 15-year-old woman raped and beaten to death on the Middle Passage and thrown overboard. Likewise with the English commoners who furiously attacked the first coal-mines opened on their ancestral lands, and the thousands of activists who protested the expansion of coal by entering mines in eastern Germany in the summer of 2016, seizing the equipment and blocking all work: they were also human beings, giving their distinctive contributions to "the totality of human actions over time" to which Ghosh refers. The foundation for this view appears to be some doctrine of original sin—although the equivalent of eating the apple has yet to be specified: but perhaps just becoming human was enough—or some hyper-deterministic and teleological philosophy of history, according to which humans were always predestined to burn fossil fuels on a mass scale, regardless of their identity and the content of their actions. It would only be slightly more absurd to claim that every human being who has ever lived has played a part in making Donald Trump the American president. That simply isn't how history works.

Alas, Ghosh is far from alone in his view, which we may call, for want of a better term, and in a strictly descriptive sense, the *universalist* conception of climate change. It belongs to the standard fare of the Anthropocene narrative, the popular story of how humankind ascended to domination of the earth and, as the most conspicuous facet of its shared rule, destabilized the climate (Malm 2016a, 2016b; Malm and Hornborg 2014). Another Indian intellectual, Dipesh Chakrabarty, has long promoted the storyline: "The poor participate in that shared history of human evolution just as much as the rich do" (2014: 14). Since more people than ever consume more goods than ever, "the lurch into the Anthropocene has also been globally the story of some long anticipated social justice, at least in the sphere of consumption. This justice among humans, however, comes at a price" (2014: 15–16). So the poor have participated in the "evolution" of climate change no less than the rich, and justice between humans is at the source of the problem. To take but one more example, in his *Ecocriticism on the Edge: The Anthropocene as a Threshold Concept*, literary theorist Timothy Clark depicts the movements of "enormous and dense tectonic plates of humanity" that cause climate change by engaging in things like food production, poverty reduction, distribution of medical services, and, not the least, the siring of children (Clark 2015: 14—the phrase is borrowed from Michel Serres, see further e.g., 80–7, 108–11). Feeding the hungry is then no longer an unequivocal good. "The attempt to achieve just human arrangements," the editors of the journal *ariel* sum up Clark's analysis, "looks ideologically suspect on the scale of climate change" (Clarke, Halpern, and Clark 2015: 2).

Two consequences of the universalist view immediately strike the eye. First, it cannot point to radical action against fossil fuels, but rather tends to obscure the sight. If climate change is the outcome of the "very existence" of human beings—not of any particular operating procedures, but of all that we ever do and have done—it seems exceedingly difficult to imagine any other trajectory than the present. For Clark, there is no way the human species can break out of the furnace, since it is snared in "an impersonal dynamic it cannot command," the victim of its own inability to grasp the climatic effects of humdrum activities—a blindness "inherent to the [sic?] humanity per se" (2015: 149, see also 30, 90). If the poor do as much to uphold business-as-usual as the rich, there can be no target, other than humanity *in toto*. The only politics the universalist conception seems capable of inspiring is misanthropic deep ecology, whose key policy prescription was always the reduction of human numbers. Once an influential current in the green movements of the West, it is now thankfully moribund; yet another Indian scholar, Ramachandra Guha, long ago showed how its attribution of environmental woes to the most universal human behaviors "is at best irrelevant and at worst a dangerous obfuscation." It has nothing to say to the people of the Global South for whom the ecological crisis "is a question of sheer survival" (Guha 1989: 74, 81). The woman in the mud cannot gain anything from it.

Second, and as a corollary of the first, the universalist conception has few if any workable tools to offer social movements actually involved in struggles for environmental justice, in particular of the climatic variety. This is the one sphere of climate politics that has remained consistently impervious to the grand species narrative: for the activists trying to stop coal and oil from leaving the ground, opposing new highways and runways, calling on investors to divest from fossil fuels, demanding repayment of the climate debt or putting their lives on the line to save rainforests, it just makes no sense to blame global warming on all humans. The scholars who peddle that story imprison themselves in their ivory towers and look down on a species whose daily infighting they cannot join. But if there ever was a time for committed, activist, indeed militant scholarship, and if there ever was a moment when radical action against fossil fuels should be the top priority for those who study the human dimensions of the problem, it surely is now. Don't speculate on the fallen nature of humankind: enter the fray.

Most important from a scientific perspective, however, is the empirical and analytical erroneousness of the universalist view. It remains strangely aloof from the facts of the matter. One study has traced 63 percent of the industrial emissions of carbon dioxide and methane made over the period 1751–2010 to no more than 90 corporations, of the ExxonMobil and Shell type, profiting from the business of extracting fossil fuels and delivering them to the fireplaces (Heede 2014). Another found that the OECD countries put up 86 of the 107 parts per million by which the CO_2 concentration rose from 1850 to 2006 (Ciais et al. 2013). As of the year 2000, the advanced capitalist countries of the North held 16.6 percent of the world population, but were responsible for 77.1 percent of the CO_2 pumped out since 1850; the share of the US alone stood at 27.6 percent, while Nigeria had a paltry 0.2 percent, Turkey 0.5 percent, Indonesia 0.6 percent, Brazil 0.9 percent—these being countries with a historical responsibility sufficiently large to make it on a top-20 list. Most left even smaller marks. Peru, Bangladesh, and Egypt were all among them (den Elzen, Olivier, Höhne, and Janssens-Maenhout 2013).

How do we conceptualize such massively uneven responsibility? Chakrabarty tries to save his idea of justice as a bane by inventing another world: "Imagine," he writes, the counterfactual reality of a more evenly prosperous and just world made up of the same number of people and based on exploitation of cheap energy sourced from fossil fuel. Such a world would undoubtedly be more egalitarian and just—at least in terms of distribution of income and wealth—but the climate crisis would be worse! (Chakrabarty 2014: 11)

Now that is one way of turning reality inside out. What Chakrabarty here ignores is the circumstance that if all 7.5 billion human beings currently inhabiting the earth were to attain, say, the average American level of fossil energy consumption, global emissions would be more than three times higher than they are at present. What is left of the carbon budget for reaching the 2 degrees target would, in the best case, be consumed in five years, and then we would continue towards 4 and 6 and 8 degrees in the very near future. Moreover, if all of those 7.5 billion people flew twice a week, owned three villas heated by coal and had at least two SUVs parked at each one of them—that is, if they were safely ensconced in life-styles now only found at the highest income echelons—the atmosphere of the earth would soon approach that of Venus, meaning that it cannot happen. Egalitarianism on the basis of extreme affluence is a biophysical impossibility. Chakrabarty's thought experiment conceals the brutal real-ity of actually existing climate change: *some people have seized hold of the atmosphere, releasing so much carbon dioxide as to initiate the global warming from which others suffer and, at the very same time, precluding a generaliza-tion of their own consumption patterns.* Those who live inside the gala of fossil energy gluttony wreck the lives of others, the unfortunate outsiders who—for reasons of natural limits—will be forever barred from entering.

And this is very much a case of injustice. If all humans are equal—common sense universalism, in the normative understanding of the word—they have, in principle, the exact same right to the atmosphere as a sink for carbon dioxide and other greenhouse gases. The atmosphere is where such gases end up, and it can hold only a certain amount before the earth catches fire; it is a dumping site for the effluents from fossil fuel combustion, a de facto common for humanity as a whole. But some have grabbed far more than their fair share. In a research project within EJOLT (Environmental Justice Organizations, Liabilities and Trade), a coalition of environmental justice movements and progressive academ-ics in the field of political ecology, one of us has conceptualized this process as *carbon sink appropriation* (Warlenius 2016, 2017; Warlenius, Pierce, and Ramasar 2015). Some people have in fact, if not in intention, appropriated the atmosphere—colonized it even—by using so much of its sink capacity that others are blocked from ever claiming a similar or even minimally fair portion. The privileged appropriators have thereby committed an injustice, which materializes as the never-ending series of climate change impacts: torrential downpours, drought, salinization of agricultural land, and all the rest that is in the pipeline. Granted, more people can squeeze themselves into the gala—for instance, the 5 percent richest Chinese have succeeded in soaking their lives in fossil fuels, while the majority of the country's population remains stranded at very modest emissions levels—but that process is inherently exclusionary and selectively injurious (Wiedenhofer et al. 2017). Every new batch of mega-emitters achieves two things: the CO_2 concentration in the atmosphere increases further, so that climate change accelerates, and there is even less left of the sink to claim for others. The more the appropriation escalates, the more the overfilled sink leaks, the stronger the currents of ecological collapse that sweep away defenseless people: the heavier the injustice.

There is, furthermore, a distinction to keep in mind here: that between subsistence and luxury emissions. It is an irony that two of today's leading exponents of the universalist view hail from India, a country that suffers more from climate injustice than most and also, incidentally, gave the world the first major analysis of the problem: *Global Warming in an Unequal World: A Case of Environmental Colonialism*, a report written by Anil Agarwal and Sunita Narain, published in New Delhi in 1991 (Agarwal and Narain 1991). Agarwal and Narain there identified a qualitative ethical difference between emissions made in the pursuit of a luxury lifestyle and those made in the struggle for survival. The methane emissions embodied in the rice put on a poor family's dinner table is not morally equal to the emissions caused by a weekend shopping flight to London, even if their global warming potential would be comparable. There is an obscenity to the latter act totally missing in the former, which makes any talk of the poor as equal participants in climate change an exercise in—with Guha—obfuscation. If we need to cut back on emissions, it is the luxury type that should first be targeted for elimination.

On the other hand, one can, in principle, imagine a global level of subsistence CO_2 emissions that stays within the planetary boundaries. To be truly sustainable, in the sense of not causing more emissions than the sinks can absorb over time, then it must probably remain beneath 5 gigatons (or Gt) CO_2 per year—a ninth of the current level. If we take into account that there are also other greenhouse gases, a sustainable limit would probably rather be 3 $GtCO_2$ per year. If the carbon sinks—the atmosphere, but also the oceans, the forests, the soils—constitute a global commons to which all humans have an equal right, then all countries could emit the sum of their per capita shares of this sustainable limit without stealing anyone else's part. Countries that over time have emitted more than their per capita share can therefore be said to have built up an emissions debt, both to other countries that have emitted less than their share (the historical component of the debt) and to future generations that have to deal with the excess (the intergenerational component). Trespassing on the rights of others, and dumping on them the consequences, those gluttons owe a massive total debt.

One of us has calculated that, unsurprisingly, the United States has the largest such debt, standing at 344 $GtCO_2$ in 2011, followed at a distant second place by Russia (88 $GtCO_2$), then Germany (75), then the United Kingdom (63). The list of claimants is headed by India, with a negative historical debt of 26 $GtCO_2$, followed by Bangladesh (6.5 $GtCO_2$), Pakistan (3.7), and Ethiopia/Eritrea (3.1) (Warlenius 2017; the figures and a visualization of the data are available at https://ejatlas.org/featured/climate-debt). These results dovetail neatly with another fresh study calculating the climate debts accumulated since 1990. Again, the United States is the number one debtor, followed by Russia, Canada, Germany, and the United Kingdom, while the biggest creditors—that is, countries with large populations that have emitted less than their per capita shares—are India, alone holding a third of the credit, followed by China, Bangladesh, Pakistan, and Nigeria (Matthews 2016). With its one-billion-plus population, India is responsible for a trivial 3 percent of cumulative CO_2 emissions so far in history (Canadell and Raupach 2014; Matthews 2016). That is some number to keep in mind the next time several hundreds of millions of poor Indians suffer acute water shortages and armed guards are stationed around the dams of the country (Agence France-Presse 2016a). It will be carbon sink appropriation coming home to roost.

Much as we should expect from this theory, there are no signs that emissions levels are undergoing equalization along the lines of Chakrabarty's fantasy scenario. Subsistence emissions from the poor part of humankind make up a tiny fraction of the total: one-tenth of the species accounts for half of all present emissions from consumption, while the bottom half of the species is responsible for one-tenth. The richest 1 percent in the world have a carbon footprint some 175 times that of the poorest 10 percent; the emissions of the richest 1 percent of Americans,

Luxembourgians, and Saudi Arabians are 2,000 times larger than those of the poorest Hondurans, Mozambicans, or Rwandans. The assumptions behind those findings are conservative (Chancel and Piketty 2015; Oxfam 2015). An adherent of philosophical liberalism would perhaps reflexively object that there is nothing unjust about such inequalities: owning much is perfectly fine, as long as enough is left for others. But the fulfilment of this classical Lockean criterion for just private property is *exactly what carbon sink appropriation rules out*. Nor can the liberal belief in a trickle-down effect have any bearing on this case, since the losers will sustain catastrophic losses *in proportion to the winners' gains*—such as, for instance, being overrun by a giant mud river (Singer 2004: 28–31).

There can be no justification for carbon sink appropriation. It is a blatant form of double injustice, the victims carrying the burden of a party from which they are simultaneously excluded. The furnace is there because it has been a feast for a select few. It appears, then, that movements for climate justice and all aligned forces should demand *the decolonization of the atmosphere*. The rich colonizers should be compelled to clean up the mess they have created, slash their own emissions to zero, finance the global transition away from fossil fuels to renewable energy— at the same time eradicating energy poverty in the Global South—fully compensate for the loss and damage caused by climate change, such as extreme weather events tearing apart countries in the Global South, and begin to actively remove CO_2 from the atmosphere. That is, of course, easier said than done. But it is an agenda in line with the material interests of the wretched of this warming earth.

5 | TAKING ON BIG OIL BY LOOKING WITHIN[1]

Anjali Appadurai

The fossil fuel industry continues to flourish as the world's largest industry, endowed with immense political and economic power, firmly established as the world's primary source of energy. It became clear in the final, desperate years leading up to the much-anticipated 2015 United Nations climate negotiations at COP 21 in Paris, that governments were not able to take meaningful action as long as they continued to operate as fossil fuel-addicted economies. In the wake of this failed multilateral process to address climate change, a wide variety of communities around the world are left to bear the costs of climate impacts. As the contemporary social movement of climate and ecological justice shifts much of its collective energy away from the UN process, a critical question looms over our movement: How do we address the continued dominance of the fossil fuel industry?

In this chapter, I argue that the prevalent attitude that individual consumers are as responsible for causing and addressing climate change as companies and governments contributes hugely to maintaining the social license[2] of the fossil fuel industry. This attitude is reflected in the social and moral pressure to adopt the ever-accelerating trends of eco-consumerism and "green" lifestyle choices, as well as in corporations' attempts to lure customers with assurances that their practices are sustainable—in short, in the rise of green capitalism. Green capitalism thrives upon people's guilt over climate change and the responsibility they feel to make a difference through their spending choices. Here enters the hierarchy of access and privilege: many green alternatives require significant money, time, and other resources, which means they become more accessible to the economically privileged. And perhaps the greatest injustice of personal responsibilization and guilt is that it can isolate and distract us from collective action—which to my mind, is perhaps our best hope for addressing the climate crisis. While being a conscientious consumer is a worthy effort, it should not take the spotlight off the huge industrial players and the more systemic roots of climate change.

The imperative to hold companies responsible is made all the more difficult by the fact that oil is a constant presence in our lives, a primary source of energy for almost all our daily activities, and the raw material for many of the objects that we depend upon to live. This dependency is a moral upper hand that industry and pro-oil commentators often use to silence critique of the fossil fuel industry. My goal in this chapter is to expose this false dichotomy of "oil user vs. oil critic" and highlight the fundamental responsibility of the fossil fuel industry to own up to the tremendous costs of its products and operations.

A critical task for the climate justice movement is to draw people away from their sense of individual guilt or responsibility for climate change, and point to the historic and ongoing responsibility of the fossil fuel industry. Only then can we erode the industry's social license and help pave the way for renewable, community-owned energy, and a climate-safe future.

A global movement to address corporate climate accountability is underway, exemplified by the '#ExxonKnew' campaign and many recent municipal lawsuits brought against fossil fuel companies for their role in creating the climate crisis. This work isn't easy: the hold of these companies upon our lives goes beyond economics and energy politics. Their responsibility to pay for their contribution to the climate crisis is often shrouded by deep-seated public misconceptions about the harm caused by their products and their role in creating a fossil fuel-dependent economy.

In the face of the devastating impacts of climate change upon vulnerable nations and communities around the world, multiple human rights abuses by oil companies (Center for Constitutional Rights 2009), health impacts faced by communities in extraction sites (National Resources Defense Council 2014), and the numerous spills, leaks, and environmental disasters from oil infrastructure projects, the true cost of our fossil fuel economies looms large.

Does it help to hold ourselves responsible?

In the mid-to-late 1980s when global concern was rising over a hole in the ozone layer in the atmosphere, everyone understood that the goal was to control products that contained ozone-destroying substances such as those used in refrigeration. Similarly, efforts to phase out lead additives in gasoline focused on ensuring that companies that produced gasoline stopped putting lead into it. During these times, public debates about individual consumers' share of the ozone problem (through refrigerator use) did not manage to stall the eventual establishment of a global regime to control the industries most responsible for ozone pollution. Nor did naysayers accuse those concerned about lead pollution harming their kids of being hypocrites because they owned cars. Individuals were not seen as responsible for what was clearly an issue of unregulated corporate activity. These were systemic problems that *could not be solved* by individuals acting alone.

Many people around the world are deeply concerned about climate change. They want to do their part—and for others to do their part—to reduce their impact on climate change (Poushter 2016). In fact, we know that most people are conscientious with regard to climate change and climate action when they are aware of the facts. In affluent nations, the main public conversation about climate solutions can often center around lifestyle activism and green consumerism, which places undue judgment and guilt on the individual. It also ignores the reality that individuals are caught in a climate-unfriendly system, and can only do so much. It is a disempowering conversation because it sets up a grand problem that needs solving—climate change—and offers a solution that lacks the power to make a real difference.

When it comes to "green" alternatives and conscious consumerism, it also becomes a question of access and privilege: we cannot all afford to buy electric cars or retrofit our homes to be more energy efficient—and even if we did, would that really solve the crisis? Even smaller, conscious actions like growing a vegetable garden or biking to work are ultimately unlikely to bring about systemic change.

The role of the fossil fuel companies

In the negotiations for the *Montreal Protocol on Ozone Depleting Substances* in the 1980s, why didn't negotiations focus on a connection between individual usage of ozone-depleting chemicals, such as chlorofluorocarbons (CFCs), and the global problem? Perhaps it was because individual ozone-damaging appliances, like refrigerators, were too essential for daily life. Or perhaps it was because most people believed that industry was more responsible for solving the issue than they—as consumers—were.

In the case of the fossil fuel industry's rampant extraction, burning, and marketing of dangerous emissions, we encounter a comparable situation. The difference is that in this case, we've been misled for decades by the fossil fuel industry as it embarked on a well-funded campaign to take the spotlight off its own responsibility for climate change. Their campaigns made us believe that we, as individual citizens, bear equal responsibility for reducing emissions (Center for International Environmental Law 2016).

Individual Canadians, for example, are made to feel guilty for causing (on average) 20.6 metric tons of greenhouse gases per year (and in reality that includes a share of Canada's industrial emissions) (Environment and Climate Change Canada 2017). Meanwhile, global oil giant ExxonMobil's operations and products resulted (in 2014) in about 560,600,000 metric tons of greenhouse gas emissions. For this contribution to climate change, Exxon made US$32.5 billion (ExxonMobil 2015). In total, about 2 percent of the human-caused greenhouse gases emissions in the global atmosphere come solely from Exxon's—a single company's—operations and products (Griffin 2017).

Exxon scientists have had awareness of some the risks of fossil fuel use and climate change since at least 1957 (Brannon, Jr. et al. 1957), and the company was well aware of the risks that its products were causing by the 1970s (Banerjee, Song, and Hasemyer 2015). Since that time, the company used its considerable resources to fund public misinformation on the science of climate change (see e.g., Exxonsecrets.org n.d.; Smoke and Fumes 2016), financially support climate-denying politicians (Goldberg 2015), and while aggressively lobbying against government action. It should also be noted that the fossil fuel industry worldwide is subsidized by many world governments to the tune of US$5.3 trillion per year, according to an International Monetary Fund (IMF) estimate (2015).

This cocktail of factors—massive deception campaigns, Big Oil's deep reach into our political and economic systems, and the ubiquity of oil in our daily lives—has meant that the industry has historically never been held accountable for a fair share of the ever-rising costs of climate change. Meanwhile, these costs are being incurred anyway, and local communities around the world experiencing climate impacts are currently paying for them.

Oil flows through every corner of our global economy, embedding itself as our primary source of energy and the destruction of our environment and the climate. And yet, the burden of guilt that many face from their perceived responsibility for climate change continues to grow as we hurtle towards a disastrously warmer world.

Capitalist systems create oil dependency. Individuals can take personal and symbolic steps to reduce their own contribution to climate change. However, real systemic change can only emerge from organized, collective action, and a recognition that, in order to be heard by fossil fuel companies and governments, voices must be raised together. Together, raised voices can engage with power, and begin to take on the vested corporate interests that benefit from our fossil fuel-addicted system. True people power requires the climate justice movement to practice solidarity and unity across its incredible breadth and diversity, and present a united front against the global fossil fuel industry.

Who should pay for the costs of oil dependency?

Right now, it is communities all over the world who are paying for the costs of preparing for, and recovering from, climate impacts: local governments most often shoulder the costs of damages from climate-related weather events, or climate adaptation projects. If the main players causing climate change—the fossil fuel industry—had to pay for the true costs of the damages they cause, they would be forced to pivot away from business as usual, or perhaps go entirely bankrupt. It is estimated that climate change will cost the Canadian economy C$5 billion per year by 2020. Metro Vancouver municipalities are expected to pay C$9.5 billion for sea level rise before 2100 (Pynn 2012). Considering these examples, and others, climate change and its ever-accelerating costs would become an issue to take seriously, if only for its exorbitant financial bottom lines. Our communities will not be able to afford the rising tide of climate costs that is bearing down on them, unless we open a critical public debate about who should help pay these costs—a debate which could lead to real fossil fuel financial accountability.

It is urgent to discuss the responsibility and accountability of the industry that has played a major role in creating climate change, an industry that continues to oppose changes to the oil-addicted economy. As a result, the fossil fuel industry profits to the tune of billions of dollars from selling products which it is commonly acknowledged are destroying our atmosphere and our communities.

The rising level of public support for bold climate action like New York's 2018 lawsuit against five major oil companies shows that this type of action has hit a nerve: frustrated by the failure of the UN and governments to set policies that curb emissions, many are starting to question the role of fossil fuel companies. We have every right to demand that these companies pay a fair share of the costs of climate impacts that they had a hand in creating. This knowledge is the beginning of true climate accountability.

It's time to stop taking personal responsibility for climate change and so granting immunity to Big Oil.

Notes

1. This chapter is a revised version of a blog post for the Climate Law in Our Hands program at West Coast Environmental Law, a non-profit organization based in Vancouver, British Columbia, Canada. The Climate Law in Our Hands initiative seeks to place the spotlight upon the fossil fuel industry's historic and ongoing contribution to climate change, and highlight its failure to pay any share of the costs arising from its products.

2. Social license refers to the "ongoing approval or broad social acceptance" for a project within a local community and among other stakeholders (Shinglespit Consultants 2017).

6 | CLIMATE CHANGE FORCES POST-CAPITALISM

Kim Stanley Robinson

Editors' introduction: This chapter is based on a keynote talk that novelist Kim Stanley Robinson gave in May 2016 for "Climate Change: Views from the Humanities—a Nearly Carbon-Neutral Conference" (http://ehc.english. ucsb.edu/?page_id=12687). The talk was transcribed by Krystal Baca and edited by John Foran. The reader might compare the scenario envisioned in this talk with the plot of the novel New York 2140, *published not long after the talk was given.*

The necessity for post-capitalism has come to us because of our need to deal with climate change. We are on the verge of creating an anthropogenic mass extinction event, meaning one that *we* have caused. Climate is not the only issue involved here. We talk about climate a lot because it's something we can conceptualize and it resembles turning the thermostat up and down, it feels like something that we can actually *do* something about, but it is only one part of the problem, which is the more general environmental collapse of which climate change is one of the main symptoms.

People talk about it being easier to imagine the end of the world than the end of capitalism. It doesn't look like we can afford the future. It doesn't look like we can pay for saving the Earth. In this ideological fixation of ours, that only the free market should rule, we are hurtling off a cliff into a disaster, not just for our species but for the entire biosphere. The mass extinction event will mean extinction for many creatures that have been here for millions of years. I'm going to say that this phrase, "it's easier to imagine the end of the world than the end of capitalism," partly suggests that the end of *the world* is bad. Maybe what the phrase is saying is that the end of capitalism is good—that it's easier to imagine bad things than it is to imagine good things. It's easier in my terms, as a science fiction writer, to do dystopia than to do utopia. It certainly seems true in terms of drama and setting up narrative tension and danger for characters to imagine a bad place and do a dystopia and imagine the world ending than it is to imagine a utopian future in which things go right in a complicated world that includes the trajectories that are already in place.

We are really imaginative, and it's by no means impossible to imagine the end of capitalism—you just have to run through it in a step by step fashion. First, you can try to imagine: if things were going right, what would it be like? The second step, if that is a desirable end state, then, is: how do we get from here to there? It's not just imagining a utopian system, which is really just an experiment—a thought experiment, a thought exercise—but also trying to imagine a plausible history that gets us from here to there. I am very interested in both things. When people look at the situation within late capitalism it looks as if finance rules. Essentially capital rules, so then: what is the state, what is the nation-state? Is the state merely being bought by capital and therefore is it essentially, effectively, the police and the enforcers of the capitalist system which is highly oligarchic and not democratic? Or is the state really an expression of people at large and therefore what it says it is? Which is to interrogate democratic systems, and here you have to come to the realization that democracy and capitalism are not allied at all in the moment that we are in. In fact, there has been a concerted effort by capital to buy the governments of the world so that whatever it is people vote for is not actually what the governments do, which is always to support finance over people, always to support debt; and public debt means people debt.

If you imagine that government is "of the people, by the people, for the people," you also have to imagine that such a situation is under attack, that it can be lost, that it always has to be fought for. Obviously, it does not exist right now because we keep electing politicians in the US who go and do what the banks want them to do. And the banks are actually funding their campaigns, so as to whether they are responding to their voters or their campaign contributors, the results seem clear: they are responding to their campaign contributors and to money itself.

And then you get to a notion of the state as a battle ground, as the site of contestation, that it's not a done deal. It isn't clear what the state is, or the various definitions that are often made by anti-statists of one sort or another—libertarians, anarchists, leftists of all varieties who are somehow convinced that the state is so allied with capital that really the two are synonymous. You have to detach yourself from that point of view and think, no, actually, the state is always open for contestation. And it's a little ironic to think that seizing the state by votes alone and then changing the laws would be almost equivalent to a coup, but it would be a legal coup, in that you just follow the laws of the land by making political parties do what is good for the people and the planet rather than what's good for finance. This is a little bit of a revolution but it's also perfectly legal and uses what the cover story of actually existing democratic governments says.

What would be interesting is to take that seriously and to force the parties and the governments to do what's good for people and the planet by way of mass action, and to do this not only by voting but also, effectively, a kind of civil resistance of the people against finance. In that struggle one often hears a certain pessimism or cynicism, that "We've lost before we've begun," because the fight has been going on for a long time and since 1981, when the Reagan–Thatcher counter-revolution started the neoliberal turn, there have not been many victories, and that we have decisively lost in these last 36 years. This is true in some senses, but the battle is never quite over. If government has become a kind of a façade or a Kabuki, it would be interesting to see the Kabuki players suddenly using real knives and take over the play and change the script by forcing the government to actually be "of the people, by the people, for the people" and act in defense of future generations and the people and the environment, adequacy for all, and survivability for the other animals on the planet.

How might this work? Think about finance, with its one simple algorithm that more is better and that profit and shareholder value are the only rubrics of success. You can call this greed—not the personal greed of individuals but a system that is greedy as such. Having defined itself so simply, without any feedback loops to control what it's up to, the system is fragile; they call it efficient, like "just-in-time" manufacturing, but efficiency itself is anti-human and fragile. And economic efficiency is just a measure of how quickly money moves from everybody to the rich. So it's not actually a physical value of any use whatsoever.

What you would really want in the system would be robustness or resilience. Efficiency is not robust, nor is it resilient: when you've got just-in-time manufacturing if there is one break in that chain of events then the whole system breaks down and that's not resilience. What you would want to build into a system is resilience; modern finance capital does not have it. It is efficient but weak, fragile, and brittle because of that very efficiency. Maximizing profit and shareholder value has left it leveraged out over an abyss, leveraged in an economic sense. All of the big private banks and the big investment firms have taken their assets, borrowed off of them, loaned off of them again, and they have managed to leverage themselves out to 20–50 times as much as they actually have in hand. After the 2008 crash, Congress tried to impose some reforms on the American banks and the banks were so resistant to this that the new rules merely state that they have to have 3–5 percent of their loan amounts in assets in hand, which means if there was ever any kind of run or crash or a break in the chain of efficiencies, in terms of money coming into them, they would immediately fall apart like they did in 2008. And one of the sources of their income is people's payments, their debts, and especially mortgages, utilities payments, pension payments, student debt, and rent in general. They take in this steady stream of money from millions of people which is contractually due to them. They rely on it coming in, they assume it's going to come in as contractually obliged on a regular basis, but then they bundle it, repackage it, sell it again, loan it out, and leverage it. This flow of money coming in is absolutely necessary to have but it's also just a springboard for the creation of even more financial value. What can then happen is what happened in 2008—a sub-prime mortgage crisis. A lot of people had promised to pay money and didn't have it. When the time came for them to pay it and they couldn't, it was an accidental mass action, like a strike, where enough people defaulted on these sub-prime mortgages that the system that relied on them being good crashed, and finance had to go to the government and say "bail us out." And indeed in 2008 the government bailed them out. At 100 cents on the dollar, the banks were saved; the individual depositors and the people who failed in their mortgages were not saved. What is interesting about this is nothing changed afterward in terms of the laws. There was the Dodd-Frank Act to rein in the banks, followed by the Sarbanes-Oxley Act; these are not trivial or inconsiderable but they weren't enough to solve the problem and we are back where we started.

This is the way capitalism works when it has such a simple algorithm. The bubble bubbles up and at some point there's a collapse, things are wrecked for a while, there's a recession or a depression, and then it rebuilds. Capitalist apologists will say that's just "creative destruction," but in fact it is very destructive destruction. What's interesting, however, is that it's inevitable that it will happen again and it could be done *on purpose* as a kind of a strike against finance, a people's strike against finance. All you would need is an agreement and a plan and a government ready to take the right action as a result. A certain point comes along—July 4, 20XX—and everybody agrees that it's going to happen, everybody has the plan, maybe everyone is part of a householders' union so that you have some legal coverage, maybe it's nothing but a legal action that planned everyone going to the bank and withdrawing on the same day. People withdrawing their deposits is an impossible thing for the banks to agree to because they don't have the assets in hand to pay every one of their depositors in full, having lent it out elsewhere.

It's *possible* that citizens can therefore crash finance and cause it to go into a tailspin and then into bankruptcy where big finance can't pay each other for the massive indebtedness of their interlocked system. At that point they will go to the government again and say "we need a bail out, something has happened here." What the government would say to finance at that point is "Yes, we will bail you out, we need finance. If finance crashes then the world economy crashes and everybody hurts really badly. We'll save you, but now we own you." This would be like General Motors in 2008, where we nationalized General Motors for a while and made them pay us back for what we loaned them so they could get back on their feet.

Do *that* to the banks that failed so much more disastrously than General Motors. We gave about $7.7 trillion to the banks to keep them solvent after the 2008 crisis, a sum of money bigger than the Louisiana Purchase, the Marshall plan, the Vietnam War, the Iraq wars, and the Savings and Loan crisis of the 1980s put together. In this strike or declaration of odious debt, strategic defaulting, this mass action, this declaration of a jubilee, at that point we need the government to say "Okay, we'll save you but we are nationalizing the banks, you are now all federal credit unions." They do what banks do, they still make loans, they still make profit but that profit belongs to the US Treasury and the US people. Bankers can still make a decent salary; they don't need the grotesque bonuses that only give them poor health anyway. But if they know that they are still making a very good living, and they know that their work is suddenly patriotic and a service to the American people, they'll adjust to it. I'm not saying that capital won't fight this tooth and claw out of some instinctive notion that more is always better no matter what. As adults, people are very inflexible in their ideological positions; they believe in the religion of capitalism, so it's a little hard to make this adjustment, but they'll do it if they and their family and children will be all right personally. In any case, now they'll have to do what the law tells them to.

So at that point, you have the state controlling finance instead of finance controlling the state. It would be a major judo move, it would be highly controversial, and it would be a historical moment of the first order. But it's not magical—it's really just a set of financial actions that people could actually carry out. After nationalizing the banks, it might be the great moment to institute a Picketty tax. In *Capital in the Twenty-First Century* Thomas Picketty's (2014) main argument was that the rich get richer and the poor grow poorer, except in exceptional circumstances like World War II and immediately afterward. This is what I call "the sky is blue" science where through enormous effort and statistical rigor someone proves that the sky is blue. "The rich get richer and the poor get poorer" is one of the first phrases in *Bartlett's Quotations*; it's one of the oldest phrases on Earth. Proving this statistically is only of limited value but it is nice to have it in cold, hard, quantitative pseudo-mathematical detail by an economist. It's obvious but it needs to be said as many times as possible. Picketty (2014) said what we can do about it is to institute progressive taxation—again an old idea, and not that revolutionary—but he said it should not just be on their personal incomes, but on capital assets as well, both personal and corporate: the bigger a company was, the more it was to pay out of its asset base. If a company of a billion dollars was forced to pay 20 percent of its assets every year this would quickly downsize it and it would quickly break itself up to reduce its prominence in the progressive tax structure. Once again, government is no longer indebted to finance but in fact finds that it has an enormous amount of surplus at that point because it's saying "The people made this, by working 'of the people, by the people and for the people,' so we are saying that this is something we have to do."

There would be all kinds of secondary repercussions to something like this, and enormous turmoil. But what you can then say is that the state can afford healthcare for everybody. The state can afford free education for everybody right through college. The state could afford full employment, employing everyone who didn't have a job otherwise. That alone would be a paradigm buster because if the government offered a job to everybody then suddenly private industry would have to compete for labor and would have to pay it commensurably to the supply and demand dynamic. Private wages would rise if there was full employment, especially if the government set a minimum wage that was a living wage. Now you have a living wage for everybody, plus adequate education, and what had up till then seemed a utopian dream of "food, water, shelter, clothing, education, healthcare, and work for everybody, equally."

This is the goal. Combined with an Earth ethic of seeing that we are part of a biosphere, if people had basic adequate incomes they would stop destroying the environment; with nobody spectacularly rich or painfully poor, getting into balance with the biosphere would be the main project. There would be all kinds of new jobs in landscape restoration or manning sailboats rather than diesel boats because we would still need trade and we would still need to travel.

What's interesting is if you combine these social and economic changes with technical improvements that lead to cleaner tech, what you get is a kind of net balance: a permaculture where human civilization is actually in a sustainable, renewable balance with the biosphere itself. The biosphere is then able to give us what we need in terms of food and shelter, but also gets to renew itself and process our waste and our poisons, all the chemicals that never existed before. All of this physical flow involves work. The only new tech we'd need is what we call economics and politics, the two "technologies" that are our software. Just as computers can't run without software, technology is not just inert matter. Our technologies are systems, so you have to think of languages, of justice as a technology, of the rule of law as a technology—the software we use necessarily to keep going. Seven billion people on this planet is already dangerously over-leveraging the biosphere. It can be done but it has to be done with really smart tech, the main one being all the social systems that we use to decide and run everything else.

What I just told you is a kind of utopian science fiction story. But I would also like to point out that these practical steps are political positions that are already on the table. Some of them are already law in social democracies, others are bills that have been proposed or rights that have been asserted around the world. The sequence would start with anti-austerity, because austerity is just "The rich get richer, the poor get poorer, make people suffer, make capital have more capital, see if that solves the problem." It's actually the crazy response of doing the same thing that got us in trouble in the first place and seeing if it might work this time. Austerity is the cynical notion that "Since things won't work I need as much as I can before the whole thing falls apart." Anti-austerity means Keynesianism. All this bashing of "government" over the last 36 years has been disastrous for people and the planet. We need a return of government control of the system and of finance itself. People over capital. Keynesianism is a first step, then you get social democracies as in Scandinavia and elsewhere. This should never be forgotten: social democracies are places where you are saying that there is more than just profit and shareholder value. There are other, more overriding concerns, social concerns, in a social democracy.

After that maybe you get to a space that I would like to call post-capitalism because I don't want to weigh it down with the baggage of any earlier system that has ever been proposed or devised. If you talk about socialism—meaning more power to Bernie Sanders—it really comes down to bringing in baggage from the 20th century that might not be how things work in the high-tech 21st century. Let's just call it post-capitalism! After social democracy there will be further improvements towards what I sometimes call permaculture—because it's both permeant culture and its permutating culture. Permaculture is a good word. What you want is sustainable *civilization*, a balance of adequacy over the long haul. Whatever we call what we get to, the steps that need to be taken along the way are simply political steps that can be done legally by votes, by politicians passing bills, by people supporting those politicians and not the other ones, and by seeing the situation clearly enough to believe that there is a way forward.

In this scenario, there needs to be mass action like the civil disobedience I described—the strategic default, the deliberate crashing of the banks and taking them over. Nationalizing the banks is the key phase in the story I just told. This is not some endgame utopia, that if we just did these things everything would be all right. That's never going to happen. It's going to be dynamic and problematic for centuries to come and maybe always. You have to get over the idea that we can ever propose solutions that will hold; these are just some of the possible steps we might contemplate and one of the stories we might tell each other. These stories need to be widely shared until enough people think one of them is the right way to go and could lead to success. I offer this as one utopian science fiction story and hope it stimulates discussion and action going forward.

PART TWO

CLIMATE CHANGE THROUGH LENSES OF DIVERSITY

7 | ZOOMING IN, CALLING OUT: (M)ANTHROPOGENIC CLIMATE CHANGE THROUGH THE LENS OF GENDER

Sherilyn MacGregor

It is now customary for climate change articles to begin with ominous statements and shocking figures: a new specter haunting the globe; sea ice melting at unprecedented speeds; less than 100 years to decarbonize; 200 million climate refugees; the greatest security threat of the 21st century. Such opening gambits are so well rehearsed that it is easy to get straight to the point. This chapter takes the existence of dangerous climate change as fact and the notion that all humans are "in this mess together" as fiction. The main message will be that the myth of human homogeneity that is held up to present crises should be challenged by arguments about inequality and difference. As a result of the lobbying efforts of feminist environmental organizations, there is a growing recognition that gender ought to be added to the mix of issues considered by UN climate policy. The burden of proof, however, is most often on those who would engage in the critical practice of bringing gender into focus in the climate change picture. The default position in many circles (academic ones in particular) is gender neutrality. I begin, therefore, with two linked questions: what does it mean to view climate change *through a gendered lens* and what is gained in doing so? My discussion "calls out" gender-blindness by "zooming in" on the myriad ways that gendered power structures are visible in all aspects of climate change, from its causes to its consequences. The chapter ends with some reflections on why these questions are important for future-oriented visions of climate justice.

The construction of climate change as a global problem of humanity-at-large is part of a scientific consensus that has formed over the past several decades. The dominant approach to understanding the problem is to zoom out as far as possible on the human species and its millennia of history on the planet. For example, in defining a new geological epoch called the Anthropocene, a committee of the world's stratigraphic geologists have given an extremely broad brush to human impacts on Earth: "We are now in some ways rivaling the great forces of nature in terms of the scale of our influence on the surface of the planet" (Wing, quoted in Panko 2016). Human influence in "the Age of Man," the start of which they mark at a point somewhere in the mid-20th century, includes dramatic changes to the climate system, which is commonly referred to as anthropogenic climate change. The American Geophysical Union (2013) says "Humanity is the major influence on the global climate change observed over the past 50 years." The use of a species-totalizing Man-banner, when deployed by climate scientists and geologists, who speak the language of computer modeling and think in deep geological time, is perhaps understandable. But when these framings are adopted uncritically by environmental activists and academics working in the environmental humanities and social sciences, and lead to the presentation of climate change as a gender-neutral phenomenon (for example, see Lever-Tracy 2010), it is a source of irritation for feminist scholars. Giovanna Di Chiro (2017), for one, offers a thoroughgoing review of environmental, scientific and social scientific engagement with Anthropocene discourses, which leads her to the conclusion that feminist "contributions are largely drowned out by the torrent of 'mansplaining' (Solnit 2012a) that dominates mainstream climate science and climate politics" (2017: 488). My response to this diagnosis is that blindness to gender (and race, class/caste, sexuality, etc.) should be called out and strong, corrective lenses should be prescribed.

Gender is a contested and context-specific concept. It will be defined here as a constituent dimension of social relations that is structural in nature and based on perceived differences between the sexes; it is a primary axis of power. Because it is deeply entrenched in cultures, institutions, interpersonal relationships, and individual identities, it can be difficult to see. It can be at once omnipresent and invisible in many spheres of social life. A long-standing feminist criticism is that most scholarly inquiry is blind to gender and therefore offers flawed and partial analyses of the world (see also Haraway 1988). For well over four decades, feminist scholars have catalogued the affliction of gender-blindness that has resulted in neutral-yet-specific concepts such as "rational economic man" in economics, "the worker" in Marxist theory, and "the citizen" in political science. Feminist critiques of such blunt theoretical objects give a rationale for the application of "the lens of gender": an analytical tool that brings power and inequality into focus and theorizes gender as one of several, intersecting axes of social differentiation. It allows a selective focus on the construction and performance of gender identities, as well as how societies organize power, work, pleasure, resource distribution, and knowledge along gender lines, not only but very obviously between men and women. The lens picks up material and ideational dimensions; it illuminates the work gender does to shape and constrain the everyday lives of embodied beings as well as the images and stories that make up the common sense in human cultures. Adjusted to maximum resolution, a gendered lens magnifies the complex processes involved when people "do gender," resist gender, and queer gender in different contexts and sites. Importantly for the topic at hand, the lens of gender enables critical deconstruction of environmental problems, such as climate change, in order to analyze them as social-cultural phenomena.[1]

What does it mean to view climate change through a gendered lens? A growing body of scholarship zooms in with a gendered lens in order to bring a number of dimensions into fine-grained focus. It makes visible how gendered power relations shape the causes and the consequences of climate change. The latter has received the most attention from researchers; detailing the gender differentiated impacts has become the most persuasive "ammunition" for activists seeking to put gender on the agenda at policy meetings (MacGregor 2010; Morrow 2017). This work comes from scholars located in a range of contexts and disciplines who use different approaches to examine how the local effects of climate change are experienced by women and men. Research by feminist political ecologists has tended to zoom in tightly on the lives of women in the Global South. Most of the empirical evidence suggests that poor rural women are affected most severely by climate change because they have fewer of the skills, resources, and power needed to cope relative to men (for a review of the literature see Djoudi et al. 2016; Pearse 2017; Sultana 2014). Women tend to be more vulnerable to extreme weather conditions and their aftermath, including the need to adapt to changes in family structure and livelihood options. The tasks of feeding, cleaning, and caring for vulnerable others—the traditional work of women and girls in most societies the world over—which are already carried out under challenging conditions, become more difficult to perform. Such transformations may result in changes to who they are and what they know. Similarly, although there is much less research on them, it is also possible to see how climate-induced changes affect men by changing the conditions in which they perform their gendered roles. Fishermen can no longer fish in dried up lakes, cattle ranchers are incentivized to swap cows for coffee, and patriarchs are forced to move away from established kin groups to find work (for example, see Gonda 2017; Pease 2016; Resurrección 2017). Male migration results in more female-headed households that lack the experience, income, and social status to stand comfortably on their own two feet. In short, climate change causes social change, and that includes profound changes in gender roles and identities in the realm of everyday life. How people perceive and respond to these changes, along with their capacity to adapt to new and volatile circumstances, are deeply gendered in ways that only a theoretically informed gender analysis can reveal.

Beyond such local impacts that are material and measurable, a gendered lens enables the viewer to zoom in on aspects of climate change that are less obvious and more obscured. Examining the historic causes of—and contemporary failure to mitigate—dangerous climate change is especially effective in challenging the homogenization of humanity that happens under the Man-banner. In their critique of the approach taken by the Anthropocene Working Group, published in *Nature*, Ellis and colleagues (2016: 192) argue that the way the Group's timeline has been drawn "instil[s] a Eurocentric, elite and technocratic narrative of human engagement with our environment that is out of sync with contemporary thought in the social sciences and the humanities." While this observation may be helpful in pointing out the geologists' errors of socio-historical interpretation and false universalism, the *Nature* article authors calling for social scientific involvement in defining the Anthropocene do nothing to challenge the sense of a Global We embedded in the concept (for a sustained critique see Haraway et al. 2016). In fact they fail spectacularly to point to the elephant in the room of environmental social science: masculinism.

In one marginalized corner of the academy, there is wide acknowledgment among feminist environmental scholars that hegemonic masculinities are associated with a dominator relationship to the natural world. For example, ecofeminist scholars have been saying for decades that it is an elite masculine worldview that devalues and exploits all things feminized (i.e., women and the biophysical world) in the name of progress and profit (Merchant 1980). It is not controversial to point out that men founded, planned, implemented, and have benefited from the main drivers of both ecological crisis and social oppression: capitalism, colonialism, and modern techno-science. Gender inequality and ecological destruction are underpinned by the same logic of domination: a master mentality in Western thought that believes certain men are entitled to rule the world due to their superior capacity to reason (Plumwood 1993). A gender lens that is calibrated to see this masculinist ideology allows for a theoretical analysis of the epistemological and ideological contours of anthropocentrism in all its forms, moving the discussion safely beyond a narrow focus on impacts and vulnerability, with all its strategic pitfalls (see Arora-Jonsson 2011; MacGregor 2010, 2017).

The intersection of hegemonic masculinities and climate change is arguably the most pressing topic of examination through the lens of gender. Yet scholars working in the environmental humanities and social sciences have paid little or no attention to the material or discursive effects of the masculinity–climate nexus (for exceptions, see Hultman 2017; Pease 2016). There is a small handful of studies that attempts to calculate the disproportionate carbon footprint associated with activities that are stereotypically masculine in Western cultures, such as car driving, extreme sports, and meat eating (Anshelm and Hultman 2014; Johnsson-Latham 2007). There is a body of evidence that suggests that white men as a group are more accepting of risks and therefore less precautionary than other social groups. This phenomenon is called "the white male effect" by social psychologists (Finucane et al. 2000; McCright and Dunlap 2015).[2] In mass surveys men tend to report higher levels of knowledge about climate science, while also being more skeptical (and actually less knowledgeable) than women (McCright 2010). It is conservative white men who have also played a key role in organizing resistance to the predictions of climate science in the USA and elsewhere (McCright and Dunlap 2011; Oreskes and Conway 2011). At the other extreme from climate denialists are those predominantly white male scientists who see climate crisis as an opportunity for innovation, professional kudos, and financial reward. For example, Fleming (2017) refers to the "climate engineers who are in search of the technological fix" as modern day Baconian supermen who can harness the powers of techno-science to control the very weather (see also Buck, Gammon, and Preston 2014). Pease (2016: 26) observes "From the male scientists in the Intergovernmental Panel on Climate Change (IPCC) to the predominantly male celebrity spokesmen on climate change to the male leaders of the environmental organizations, the politics of climate change are shaped by masculinist discourses." An explicitly political process of joining the dots between specific people and environmental impacts leads feminists to hack the popular meme with this alternative slogan: "Welcome to the white m(A)nthropocene" (Di Chiro 2017; see also Raworth 2014).

How does understanding climate change through a gendered lens contribute to the task of imagining—and working towards—climate justice? Answers should be relatively simple. Only when we see the role that gender norms play in people's experiences of and responses to a changing climate, now and in the future, will we be able to ask the kinds of questions that reveal operations of power in the historical climate "story." How climate changes are affecting interpersonal gender relationships needs attention. In extreme climate stress, will women face more male violence and oppression or will men and women find ways to share labor and cooperate in decision making in order to survive? Unless such questions are asked, it is more likely than not that the status quo will be maintained, if not exacerbated. Only when we see the part played by masculinism, not only in driving the decisions to keep extracting fossil fuels and manipulating the biosphere for profit and power, but also as central to the mind-set of those who actively deny climate change is happening, will we be able to imagine the radical political changes that are needed to cure the disease of climate change rather than treat its many symptoms. After putting climate politics under a gender microscope, ecofeminist Greta Gaard argues that "climate change and first world overconsumption are produced by masculinist ideology and will not be solved by masculinist techno-science approaches" (2015: 20). Sight should lead to insight; insight should lead to more effective action towards immediate mitigation and a more inclusive understanding of climate justice.

To conclude, it is difficult to comprehend the persistence of gender-blindness when vivid images of how climate change intersects with gender injustice are so readily available. For feminist climate change critics, the task involves risking ridicule to point out the blindingly obvious. As such it brings to mind the 19th-century cautionary tale by Hans Christian Andersen, *The Emperor's New Clothes*, where a child refuses to be silenced by monarchical vanity and calls out: "but he isn't wearing anything at all!" In my 21st-century version of the story, the emperor wears the olive-drab uniform of Humanity and proclaims that those who express doubt about the causes and consequences of anthropogenic climate change will be dismissed as stupid or skeptical. But the ecofeminist critic can see with her own eyes that there is no such uniform; what parades before her is a naked human form that is posh and pink, with a paunch and a penis; he is king of the 20 percent.[3] She zooms in with her gender lens and speaks truth to power, calling out: "But there is no Global We!"

Notes

1. There is a substantial body of feminist scholarship that has theorized in this arena; a few of my "go-to" resources include: Butler (1990); Narayan and Harding (2000); Peterson and Runyan (2013).

2. A parallel might be drawn between the denial of climate risks and the hyper-masculine banking industry's failure to recognize the warning signs of the 2008 financial crisis (see Enloe 2013).

3. Many of us have been using "The 1%" to describe the global elite, and "The 99%" to refer to the rest of "us." It now seems more accurate to think in terms of the 20 percent and perhaps to stop being sloppy with analyses of class divisions (for a compelling discussion see Reeves 2017).

8 | A CULTURE-CENTERED APPROACH TO CLIMATE CHANGE ADAPTATION: INSIGHTS FROM NEW ZEALAND[1]

Debashish Munshi, Priya A. Kurian, and Sandra L. Morrison

Climate change is not merely a crisis of "nature" but is, more importantly, "a crisis of culture," the novelist-cum-scholar Amitav Ghosh (2016: 12) proclaims in *The Great Derangement: Climate Change and the Unthinkable*. Anthropogenic climate change is often seen as the biggest crisis facing humankind, although many argue that it is, at its heart, the manifestation of a larger crisis engulfing our economic, socio-cultural, and political institutions (Assadourian 2015; Ghosh 2016; Klein 2014a). Culture, for Assadourian (2015: 98), "is not simply the arts, or values, or belief systems ... Rather, [it comprises] values, beliefs, customs, traditions, symbols, norms and institutions—combining to create the overarching frames that shape how humans perceive reality." Going beyond the idea of culture as tradition or a set of habits, Raymond Williams (1960) offered his influential reading of culture as "lived experience"—a way of comprehending how people actually lived their lives. Indeed, we would argue that culture underpins the current climate crisis. Hence, harnessing the critical elements of culture-as-lived-experience is needed to confront climate change.

Creating robust policies on climate change adaptation demands a synthesis of culture-as-lived-experience with scientific knowledges and an engagement with people from a range of constituencies. This is because a variety of stakeholders—from businesses, including primary industries, spanning horticulture and dairy farming, and tourism enterprises, to vulnerable populations, including communities living in coastal, low-lying, or nature-sensitive habitats, and Indigenous peoples, face specific challenges in dealing with climate change and have disparate views on how to prioritize adaptation efforts.

It is evident that conflicting cultural perspectives and competing understandings of the implications of climate change create uncertainties in decision-making processes (Adger et al. 2009; Adger et al. 2012; see also Chapter 10, this volume). Yet, little attention has been paid to the implications for governance and to the sustainability outcomes of diverse cultural perspectives that shape citizen engagement on climate adaptation action plans.

It is in this context that we offer a sketch of a culture-centered framework of public engagement that has the potential to enhance governance on climate adaptation. At its heart, public engagement on climate change is a means for developing adaptive capacity within communities that are experiencing climate change in some form (Evans et al. 2016). These processes of engagement include the identification of public/stakeholder identities and values, deliberating among them, and building relationships. In some cases the objective of engagement is to raise a target group's perception of risk (Kahan et al. 2012; Sheppard et al. 2011) while in other cases it is to increase such a group's ownership of an adaptation project (Bardsley and Rogers 2010). Regardless of the objective, inclusive engagement processes that are cognizant of the range of values and needs of different stakeholders and target groups, and acknowledge power dynamics, are more likely to be successful.

Engagement at deeper, deliberative levels allows policy makers to involve publics/stakeholders in decision-making on a cultural level—not simply on an administrative level—by recognizing their distinct identities in specific local contexts and building meaningful partnerships (Fresque-Baxter and Armitage 2012; Leonard, Parsons, Olawsky, and Kofod 2013). As Bardsley and Rogers (2010: 13) point out, localized adaptation strategies will only be successful "if new approaches for learning are applied and tested in partnerships to allow for explorations of effective paths to adaptation within specific biophysical and sociocultural contexts." Scientific knowledge such as data from climate science modeling exercises, for example, can best be understood and acted upon by local communities if this knowledge speaks to their unique needs and priorities. Given that different communities will experience climate change in different ways—e.g., threats to lives, houses, crops, infrastructure, and social and cultural practices from sea level rise, storms, floods, fires, droughts, and temperature rise—it is essential that adaptation strategies emerge from community discussions and actions that raise awareness of, and are grounded in, the socio-ecological context. Deliberative forums are, therefore, particularly useful in the formulation of such diverse and contextual array of adaptation strategies.

There are, however, a number of barriers to engagement. Top-down governance structures can impede public engagement, especially if a particular public feels that its values will not be acknowledged in decision-making (Paschen and Ison 2014). Such situations create a sense of distrust in public institutions among communities. This sense of distrust is particularly pronounced in vulnerable communities that see government agencies ignoring or bypassing the specific challenges they face. As Leonard et al. (2013) say, any engagement program should be mindful of the interaction and potential conflict between governance structures such as in the case of Indigenous communities and natural resource management. The Miriwoong people in Western Australia, for example, are skeptical of "externally driven adaptation options" put forward by the government given the widespread settler society practices of mining, irrigated agriculture, and the damming of rivers (Leonard et al. 2013: 630) that have had devastating environmental and social consequences. Similarly, in New Zealand, despite a formal commitment by the state to uphold the Treaty of Waitangi, it has taken years of struggle to get Māori worldviews and values accepted into newly formed arrangements of co-governance of natural resources such as the Waikato River (Payne, V. 2016).

Vulnerability and culture

The concept of "vulnerability" is also central for discussions about climate change. Yet there are many types of climate vulnerability. As Gajjar, Jain, Michael, and Singh (Chapter 24, this volume) point out, "vulnerability to climate change is socially differentiated" and state-driven policies on climate change adaptation do not always correspond to the specific needs of different groups of people. In some cases, researchers have made a case for the incorporation of traditional ecological knowledge (TEK) (Bardsley and Rogers 2010). More recently, however, climate change adaptation measures are beginning to acknowledge that vulnerability needs to be assessed not just in biophysical and economic terms but also in social and cultural terms. In incorporating "subjective values" that are not readily quantified by scientific and economic measures (Albizua and Zografos 2014; O'Brien and Wolf 2010), scholars are beginning to recognize that "something greater than money is at stake" (O'Brien and Wolf 2010: 233). For example, climate-induced rising sea levels repeatedly cut off access to a venerated *marae* (a meeting place used by Māori for religious and social gatherings) during high tides at Marokopa on the west coast of New Zealand's North Island. When this happens, members of the local *iwi* (tribe) find it difficult to attend funeral ceremonies there that are so important to the cultural fabric of the community. As O'Brien and Wolf

(2010: 233) say: "what is considered legitimate and successful adaptation depends on what people perceive to be worth preserving and achieving, including their culture and identity."

There is already a significant literature on the engagement of Indigenous peoples with climate change adaptation (Herman 2016; Jacob, McDaniels, and Hinch 2010; Leonard et al. 2013; Petheram, Campbell, High, and Stacey 2010; Sakakibara 2017; Williams and Hardison 2013; Wisner 2010), and their vulnerability in the context of climate change. Displacement of people due to climate-induced events has a major impact on Indigenous people's place-identity or the link between them and their environment (Jacob et al. 2010; Leonard et al. 2013; Sakakibara 2017). Māori in New Zealand, for example, who have a strong connection with the land on which their ancestors lived find the idea of forced relocation deeply disturbing. Aside from physical displacement, climate change can also lead to the loss of traditional knowledge and customs that are tied to ecosystem services or ecological systems (Jacob et al. 2010; Leonard et al. 2013; Sakakibara 2017; Fresque-Baxter and Armitage 2012), making them even more vulnerable.

A lack of awareness of cultural factors among policy makers and policy implementation agencies can exacerbate climate vulnerability among certain communities because of either a lack of trust in national or regional-level bureaucracies (Paschen and Ison 2014) or inadequate communication channels between decision-makers and those living and working at the frontlines of climate change (Petheram et al. 2010). It is also the case that decision-makers, when they do pay attention to culture, are often seen as viewing culture through a colonial lens; they also, frequently, characterize local customs and practices as barriers to progress on climate change (Wisner 2010). Yet, it is clear that TEK and the cultural values subsumed in it serve "as a coping mechanism for the people to process climate change" (Sakakibara 2017: 162) and TEK could be the foundation for building adaptation strategies. In addition, drawing on TEK should not mean appropriating TEK. As Williams and Hardison (2013: 542) caution, researchers as well as policy planners need to make sure that there is no unethical or unjust transfer of Indigenous knowledge: "[T]he exchange of traditional knowledge involves cultural values, multiple legal jurisdictions, risks to cultural sustainability and survival and rights to self-governance." They emphasize that decision-makers should be aware of the exact nature of the exchange and should be able to determine whether the TEK is bound by Indigenous legal rights or whether it is freely transferable before deploying it.

This is not to say that only Indigenous peoples are culturally affected by climate change, because all public/stakeholder identities are closely linked to place (Fresque-Baxter and Armitage 2012). Culture as lived experience is central to understanding how climate change affects different groups of people differentially, and, consequently, that adaptation strategies cannot be designed on a one-size-fits-all formula. When such a formula is relied upon, consciously or otherwise, it inevitably prioritizes some cultural values over others, thereby reinforcing inequality. For example, the state of a community's well-being, including how it is manifested in poverty levels, health concerns, and social isolation has a bearing on its ability to adapt to climate change (Fresque-Baxter and Armitage 2012; Petheram et al. 2010). In sum, public engagement on climate adaptation needs to emerge from an interplay among science, economics, society, politics, and culture.

We now turn to the specific context of New Zealand, an island nation in the South Pacific, where public engagement on climate change is caught up in complex interwoven tensions that are grounded fundamentally in culture. These tensions involve the economic imperatives of methane-spewing dairy intensification; the environmental aspirations of a nation that seeks to promote itself as a clean and green tourist destination; the quest of its young citizens for a better and more secure future; and the struggle of its Indigenous Māori communities to safeguard their cultural and customary rights.

The New Zealand context

As in other parts of the world, there is some official recognition of the threats of climate change to life and property in New Zealand, as well as of the need to draw on scientific advances in climate research to adapt to these rapidly changing times. Confronting climate change is part of a National Science Challenge[2] in the country. The mission of the "Deep South Challenge" is "to enable New Zealanders to adapt, manage risk, and thrive in a changing climate." The challenge aims to achieve this "through a framework that connects society with scientists through five inter-linked programmes"—Earth System Modelling and Prediction with an emphasis on improving projections of climate change; Processes and Observations with a focus on getting a better understanding of the climate system through detailed observations of climatic processes; Impacts and Implications to identify the specific risks faced by the country and direct planning on climate change adaptation; Vision Mātauranga to strengthen the capacity and capability of Māori to deal with and adapt to climate change; and Engagement to encourage citizens to make informed decisions based on climate science (Deep South Challenge 2017).

In line with the growing urgency, research on climate adaptation in New Zealand has been picking up fast. Some of the recent research has focused more specifically on how the tourism industry (Hughey and Becken 2014), the dairy industry (Kalaugher, Bornman, Clark, and Beukes 2013), and the insurance industry (Storey et al. 2017) are adapting to, or facing challenges in adapting to, climate change. There have also been research-based initiatives such as the "Making It Work" approach to community engagement on climate adaptation by New Zealand's National Institute for Water and Atmospheric Research (NIWA 2011) as well as the Ministry for the Environment's (MfE 2014) New Zealand Framework for Adapting to Climate Change.

Despite these efforts, progress on climate change adaptation has been slow due to a mismatch between the scientific evidence on the implications of climate change, inadequate response from central and local governments, and public understandings of how to respond. Manning, Lawrence, King, and Chapman (2015), for example, argue that climate change adaptation in New Zealand requires a framework which can not only overcome structural inertia in climate governance but is also sensitive to socioeconomic inequalities in the country. Similarly, Russell, Greenaway, Carswell, and Weaver (2014: 783) call for steps that go beyond a technocratic approach and incorporate the "use of appropriate languages, cultural reference points, and metaphors embedded in diverse histories of climates." In our search for cultural reference points, we have started with a group that is not only among the most vulnerable in the context of climate change, but who also hold a pivotal place as the *tangata whenua* or first peoples of Aotearoa New Zealand.

Māori cultural perspectives

Māori understandings of the world can be seen in the stories of creation that affirm that all living entities are children of Ranginui and Papatūānuku or Sky Father and Earth Mother. From this perspective, nature, the animals, plants, rivers, mountains, and lakes are all related, with each requiring respect as one would accord a fellow kin. These powerful narratives allow us, as humans, to reflect on our responsibility to other entities of nature, and understand our place and role in the universe.

The complex understandings that shape Māori perspectives on nature have profound implications for policy making on sustainability.

For example, the Māori cultural value of *kaitiakitanga* (guardianship) is transferred from generation to generation. Rather than a fixed understanding of what the *kaitiaki* (guardian) role entails, there is recognition that while the responsibility is the same, the issues and challenges change depending on the time and context of that generation. *Kaitiakitanga*

> is an inherent obligation we have to our *tupuna* [ancestors] and to our *mokopuna* [descendants]; an obligation to safeguard and care for the environment for future generations. It is a link between the past and the future, the old and the new, between the *taonga* [treasure] of the natural environment and *tangata whenua* [Indigenous people of the land]. (Selby, Mulholland, and Moore 2010)

Translating these concepts into an architecture for public engagement calls for a Kaupapa Māori approach (see Munshi, Kurian, Morrison, and Morrison 2016), a philosophical and methodological approach that validates Māori knowledge, language, and culture.

Māori trace their *whakapapa* or genealogy to the mountain and river of the place they are from, seeing themselves as inextricably part of the earth, which shapes their *kaitiaki* relationship with nature. Their role as guardians is interwoven, in turn, with the notion of *whanaungatanga* – people's relationships with *whanau* (encompassing their family and community, and all living things). Indeed, the idea of *whanau* or extended family captures the sense of multi-generational networks of people and other species with whom we are intimately connected. *Mauri* (life force) represents the integrity of all things in nature. Consequently, a fundamental responsibility of all Māori is to ensure the restoration of the *mauri* of any part of nature such as rivers, mountains, or land damaged or harmed by human activity (Morrison and Vaioleti 2012: 14). In thinking about climate change and successful adaptation to the already changing environment, a dilemma confronting scientists and governments in New Zealand, and elsewhere, has been the disconnect between the findings of scientific modeling that demonstrate the often terrifying realities of climate change and public understandings and appreciation of such changes. A key insight that comes from our culture-centered approach to climate change adaptation is the importance of changing the narratives of daily lives to reflect new realities. In the case of Māori, for example, this involves retelling the stories of the past, embedded in everyday practices of fishing, food gathering, and cultivating crops, to facilitate adaptation to the changing social and environmental contexts. The challenge is to adapt the cultural knowledge of the past to tell a different story contextualized to the present in order to prepare the community for a climate-changed future that is already here.

One fascinating retelling of an old legend involves a female *taniwha* (*taniwha* are supernatural creatures in Māori legends who lived in or near water) called Huriawa, who lives in the Te Waikoropupu spring of Golden Bay in the South Island, and is responsible for the aquifers. In the legends of the past, when the water of the aquifer changed through storms or floods, it was Huriawa's job to make the aquifer's water fresh again. Because there is recognition of the threat of climate change to the aquifers, myth-keepers and storytellers are asking what new story Huriawa will have to tell to prepare communities for climate change. The adaptation story that they are working on now incorporates an advanced warning to help communities prepare for climate change. The *taniwha*'s journey now goes from the top of the South Island where she has traditionally lived, out to sea and to many other places in the island. In the process, she will be covering several *hapu* (sub-tribal) regions and they might have a different story per region. These fluid narratives may shift as Huriawa travels, adapting the traditions of each region to tell its own story. Stories thus become mobile, shifting shape and space, to become conduits for traditional cultural champions who are still guardians of the environment, but now take on new roles to prepare their communities for adapting to new realities (Morrison 2016). What we see in such repurposing of stories and myths is the coming together of science and traditional knowledges to create a cultural context open to climate change adaptation.

At a more fundamental level, engaging with cultural constructs in climate change adaptation policies requires us to ask what does being good *kaitiaki* (guardians) mean, and entail. A culture-centered architecture of public engagement in the New Zealand context recognizes that values are grounded in *whakapapa* (genealogy), which links people to place. *Kaitiakitanga* (guardianship) leads to recognizing that it takes *whanaungatanga* (a complex of relationships that includes kinship between and among both humans and non-humans) to serve as good guardians of the environment. It also means acknowledging that every living thing has a *mauri*. Therefore, preparing for climate adaptation requires upholding the integrity of the *mauri* of everything in the world around us. For Māori, recognizing the realities of climate change also means asking: How will climate change affect the *mauri*?

These concerns come into play as Māori, especially those on the coast, consider movement away from their traditional lands. Those who already are at risk of rising sea levels, face questions that go to the core of their identity. Can their connection to place be recreated in new areas? How do we make it so that there's a river and mountain when they say their *pepeha*: their introductions that draw down their genealogy from their mountain and their river? Māori sense of place and hence identity is affected in irrevocable ways by climate change. But in a very material sense, climate change also affects their access to water and food. Maintaining and sustaining links between the cultural and the material call for the intertwining of *matauranga* knowledge with scientific insights.

In conclusion, we argue that a focus on culture is imperative in blending the advances in science and technology with local knowledge and life experiences. Indeed, centering culture in public engagement processes on climate change, grounded in insights gleaned from science and traditional knowledge, can facilitate climate adaptation among those most at risk from climate change. Such adaptation strategies work to build resilience in local communities facing climate change, and also form one part of the large-scale efforts required to cultivate "cultures of sustainability" (Assadourian 2015: 104). It is through drawing on some of the many differing meanings of culture by which we can shift power away from institutional elites and to communities on the ground to deal with the challenges of a changing climate.

Notes

1. This chapter comes out of research funded by the University of Waikato and the Deep South National Science Challenge in New Zealand. We thank Timothy McGiven for exemplary research assistance in the early stages of this project.

2. The New Zealand government, drawing on input from the scientific community and the general public, has identified 11 important challenges facing New Zealanders, which are the focus of research investment (Science Learning Hub 2016).

9 | EXORBITANT RESPONSIBILITY: GEOGRAPHIES OF CLIMATE JUSTICE

Nigel Clark and Yasmin Gunaratnam

Connecting with climate change

"If climate change makes our country uninhabitable, we will march with our wet feet into your living rooms." With this impassioned intervention at a 1995 Berlin climate change forum, Bangladeshi representative Atiq Rahman vented his frustration with the stalling of international climate negotiations (cited in Roberts and Parks 2007: 2). As burgeoning studies around the question of climate migration have since made clear, there is no simple, linear relation between vulnerability to extreme weather and the long-distance mobilization of a caring and just response. But that most likely wasn't Rahman's point.

One of the great challenges of climate change is that the scientific evidence upon which issue formation depends "cuts against the grain of ordinary human experience" (Jasanoff 2010: 237). Both its causes and effects seem too widely distributed in space and time for us to grasp palpably, immediately, and personally. "(A)s a global phenomenon, climate change is often not locally observable or easy to reconcile with laypersons' local experiences, making its seriousness sometimes challenging to convey," Grasswick writes (2014: 542). What Rahman seems to be doing, in this regard, is trying to shift the issue of climate change away from planetary modeling and abstract knowledge. He reminds us that those on the sharp edge of climate change are flesh and blood people whose potential suffering ought to be felt closer to home.

This is the kind of work that critical human geographers take to be vital and urgent. As we will see, geographers specialize in tracing the complex patterns of interconnectivity that implicate the lives of people *here* with others near or far. Though we go about this systematically— gathering as much evidence as we can—most of us take on such tasks because we care about the unequal and unjust ways that life chances are distributed in the contemporary world. But this is complicated work. Can we assume that the tracking and calculating of unfair exchanges is an effective way to make people care more about distant others? Is a calculus of trans-global gains and losses really the best means of encouraging compassion for lives very different from our own? And what about the challenges of care ethics, where a concern for others makes us vulnerable (van Dooren 2014), not least because "our recipients of care, can answer back" (Puig de la Bellacasa 2012: 209)?

Increasingly, human geographers are interested not only in the social processes that render global "playing fields" uneven, but also in the many nonhuman phenomena that help compose these bumpy, irregular realities. But things get even trickier when we factor in the workings of the earth itself. Global climate is an immensely complex system, with more connections, nodes, and feedbacks than almost any known system. In such a world, no single climatic event can be unambiguously attributed to anthropogenic influences, let alone pinned to the actions of a single group or category of people. And even if we could somehow level the global socio-economic playing field, this is a planet whose ordinary, ongoing instability would still make social life—from time to time—immensely challenging.

If it's not easy to unequivocally map out chains of causality for climate change, so too is it difficult to predict how, and under what conditions, different collectivities will react to shifting or extreme climatic conditions. For example, we have indeed recently witnessed groups of South Asians, many of whom were of Bangladeshi heritage, marching wet-footed through the streets and living rooms of Lancashire, Cumbria, and southern Scotland. They came not as displaced people, but as emergency relief squads responding to floods in the wake of Storm Desmond of December 2015. Muslim civil society organizations in the north of England—experienced in responding to extreme events overseas—mobilized quickly to provide food, supplies, and clean-up assistance to flood-struck communities (York 2015).

It is unlikely that these people came to help out of a sense of causal connection or liability. These were also communities under pressure, facing stigmatization at a time when Muslim collectivities were vilified in the media and by far right groups as "terrorists" and "extremists," and during campaigning for a referendum to leave the European Union (in June 2016) that was characterized by racism and xenophobia. They may not have felt, in advance of the floods, particularly connected to the afflicted communities, though in the act of assuming responsibility, they certainly made new connections. It is also impossible to overlook the specific context of such giving; its adjacent and intimate entanglements with a possible resistance to stigmatization. Indeed, some of these stories came to light because they were publicized on social media by the groups themselves, using hashtags such as #MuslimsForHumanity (York 2015). To what extent then, are such circumscribed gestures—a reaching out by strangers to those in need—exceptional? Or is there something rather more ordinary about such a response? What if a "geographical" imagining of justice and responsibility in a time of climate change were to set out from such situated overtures? Where might we end up? On what kind of journey might it take us?

Mapping climate injustice

In the earlier days of concern over climate change, climatologists seemed to work under the assumption that providing relevant data would be enough to spur decision-makers to deal with the problem. When evidence of present and predicted climate change—even potentially catastrophic shifts—proved insufficient to spark the necessary policy responses, it became apparent that there was more at stake than "a deficit of understanding" (Clark 2015: 160). Attracted to the irrupting debate, critical social thinkers and activists set out to show how existing patterns of energy use were bound up with powerful vested interests. They also assembled evidence that demonstrated how vulnerability to climate change mapped uncannily onto existing disadvantages associated with the vast socio-structural inequalities rifting the global economy. Already underprivileged regions are disproportionately susceptible to changing climate, especially with regard to projected agricultural outputs. Research also suggests they would find themselves under-resourced when it came to adapting to changing conditions and further disadvantaged in their efforts to maintain a strong presence in global climate negotiations (Newell 2005; Clark, Chhotray, and Few 2013).

There is a deep-seated moral-political dimension to this sort of articulation of global climate injustice. Like Rahman's outcry, such interventions not only seek to expose the inequity structured into global social orders, but attempt to bring climate change controversies back to the scale and experiences of daily life (Neimanis and Loewen Walker 2014). In this way, critical social researchers hope to add a vital charge of care and compassion to the too often self-serving and conditional world of international climate negotiation—to help jolt it out of its costly stalemates, delays, and deferrals (Roberts and Parks 2006: 221–6).

It is here that critical human geographers like to feel that our spatial imaginations and skills are especially valuable, for we see ourselves as geared up to map out the routes, vectors, and networks through which everyday lives *here* connect with lives elsewhere and to show how these pathways serve as the very medium through which unfairness is perpetuated. In this way, geographers reveal how it is that those of us living in more privileged places benefit from unequal spatial relations—in quite mundane ways. Whether it is by using oil extracted from distant lands, consuming cheap calories others have grown, or adding disproportionately to greenhouse gas emissions, our stories indicate, those of us enjoying relatively high standards of living are implicated in the underprivilege, expropriation, and suffering that is happening *elsewhere*—beyond our usual sightlines.

The assumption underlying such accounts is that by attending closely to the ways that our lives are entangled with other lives, human and nonhuman, we will feel obliged to take greater responsibility for our daily deeds and for the very organization of our interchanges with others. But lately, geographers have begun to ask themselves some tough questions about this supposed passage from recognizing causal links between *our* actions and *their* predicaments to the emergence of more caring and compassionate ways of relating. For just as overcoming "deficits" of scientific understanding about climate change does not automatically produce effective policy, neither does it appear that exposing deficits of political understanding, feeling, or of moral sensibility leads straightforwardly to appropriately virtuous dispositions or measures.

As Clive Barnett and David Land put it: "the mere fact of being bound into relationships with distant others does not actually provide any compelling reason that could account for or motivate relationships of care, concern, or obligation" (2007: 1069). Climate change is a good example. Given that anyone's personal contribution to greenhouse gas build-up will rebound through the unfathomably complex interconnectivities of the entire earth system—this would seem a rather convoluted way to come to care passionately about actual, flesh and blood people. For sure, having a reasonable sense of the mutual implication between places near and far does no harm, and indeed has become a significant part of global climate negotiation. But some critical spatial thinkers are asking whether there might not be better ways of understanding—and encouraging—the emergence of responsible and caring dispositions towards "others."

For a start, whatever news media tell us, kindness and generosity are not necessarily in short supply. As ethical thinkers point out, while they may not get the credit they deserve in competitive economies or bureaucratic systems, such virtues are the ordinary and ubiquitous "load-bearing structures of society" (Vaughan 2002: 98). Mostly murmuring away in innumerable uncelebrated acts, caring and generous overtures often flare into visibility in times of crisis, such as during Storm Desmond, Hurricane Katrina, or any number of well-documented calamities (Clark 2011). While help from those in the vicinity may be most urgently needed, there is plentiful evidence—embodied in donations, volunteering, professional organizations, and social movements—that caring gestures reach far across the planet. Moreover, such outpourings of empathy, support, and assistance in the face of extremity suggest that compassion does not wait for the revealing of causal connections or culpability.

But what happens when climate change arrives not in calamitous, rapid onset events, but in the "slower violence" of chronic environmental change or ever more routinized conditions of extremity? There has recently been growing attention to the way urban populations are responding to climate stress, especially in cities where infrastructure cannot be relied upon to cope with escalating pressure. In Mumbai, monsoon flooding is now considered "normal," while parts of Jakarta were inundated five times in 2015. In these "ordinary cities" of the Global South, numerous forms of improvised response to enhanced climatic variability can be observed, ranging from architectural innovations including green shading to reduce heat stress and the elevation of furniture or housing (Banks, Roy, and Hulme 2011), through to new social media platforms such as Jakarta's real-time flood mapping application that enables citizens to collaborate in the management of flooded cityscapes (Holderness and Turpin 2016).

Here too, on the frontline of global climate change, the question of care, compassion and generosity as everyday social "load-bearing structures" calls for special consideration. In cities of the South, associations of neighbors, relatives, and friends provide vital support for weathering extreme events—in the form of provisions, temporary shelter, information, and financial assistance (Roy, Hulme, and Jahan 2013; Jabeen, Johnson, and Allen 2010). While most researchers stress the gradual, mutual building of trust in such informal networks, some have also noted how spontaneous offerings of assistance often precede—or exceed—any reciprocal arrangement. Jonathan Shapiro Anjaria (2006), for example, provides a moving account of responses to Mumbai's exceptionally severe flooding of 2005 in which some of the city's poorest and most marginalized people effectively self-organized to help stranded strangers.

Just as we should not presume that those with relatively fewer resources will be slow to make generous offerings, neither should we assume such openings are restricted to the local scale. When super-typhoon Haiyan (Yolanda) struck the Philippines in 2013, there were once again many ground-level "stories of hope, courage, creativity, and empowerment" as low-income, under-resourced people came together to endure catastrophic conditions (Valerio 2014: 156). A great many of those who rallied to raise funds for relief and reconstruction were transnational workers. As Cleovi Mosuela and Denise Matias observe: "cross-border migrants ... constitute an international network of Filipinos who are instrumental not only in keeping the Philippine economy afloat but also in constituting a network that may serve as a response to major environmental disasters in the Philippines" (2014: 8).

Rather than supposing that we need to begin with carefully computed geographies of who owes what to whom, then, a case might be made for setting out from the mundane reality of people reaching out to each other in times of stress and need—and working up from there. If justice is going to work, to push through the barriers that are endlessly thrown up in its path, we must truly, deeply desire that others be relieved of their suffering and deprivation. But while justice may need care and compassion, these virtues themselves tend not to await a calculus of costs, debts, liabilities. They seem most often to emerge from actual encounters with others (which doesn't mean that they have to be direct or unmediated) (see Barnett and Land 2007). As we suggest in the final section, a consideration of these at once ordinary and extraordinary acts of care and compassion might help us come to terms with living on an inherently changeable planet. Though if we wish to respond both fairly and effectively to climate change, this by no means absolves us from doing the most exacting calculations.

Exorbitant responsibility

To pursue climate justice is to ask what kind of social world we inhabit—to probe its ruptures, imbalances, clashes. While we have been suggesting that tallying gains and losses might not be the only or best starting point for responding care-fully to climate change, it is also vital to recognize that any acknowledgment of widespread capacities for self-help or self-organization in a profoundly uneven world runs the risk of abetting those who would leave the poor, marginalized, and vulnerable to their own devices in times of extremity. More disturbingly, it could play into recent policy moves to encourage the selective uptake of the most flexible and resourceful disadvantaged people on the frontline of climate change into global economies—in a kind of "positive" climate migration that leaves the most vulnerable behind and takes away those who might have been best able to care for them (Bettini 2014).

In short, if we are to care more—and if we wish to help others in their own caring practices—then we also need to keep a close eye on the deeply, cruelly imbalanced forms of calculation that are already at work in the world. And this in turn is part of a more general lesson—which is that if genuine offers of assistance are to be truly effective—however much they precede or break out of economies where values are known in advance—it is necessary to intervene as knowledgeably and judiciously as we possibly can.

As philosopher Jacques Derrida asserts, "one can't make a responsible decision ... without knowing what one is doing, for what reasons, in view of what and under what conditions" (1995: 24). This means that whenever we make a gift or add our weight to a political conflict, we should also accept that our offerings are quite likely to fall short or miss the mark. To recognize, therefore, that overtures of care or struggles for justice are inevitably learning processes in which we find ourselves interrogated, provoked, inspired by the singularity and specific needs of those to whom we attend (Gunaratnam 2013: 47–50).

In the case of climate change, such attempts to figure out "what we are doing" not only means that we ask what kind of *social* world we live in, but *what kind of planet* we inhabit. Over the last 50 years, Western science has offered ever more evidence of the inherent dynamism of the earth—conveyed most dramatically in theories of abrupt climate change and the Anthropocene thesis. As geoscientists insist, "detailed paleo-records show that the Earth is never static ... variability abounds at nearly all spatial and temporal scales" (Steffen et al. 2004: 295). For many peoples whose cultural memories and earth stories cover long periods, however, such scientific revelations are unlikely to come as a surprise. Here, ecological, geophysical, and Indigenous knowledge is often bound up with practices, values, and ways of relating that are deeply oriented to the varying demands of a profoundly changeable world.

In this sense, a crucial aspect of pursuing and enacting climate justice would be acknowledging, learning from, "doing justice" to, the hard-won achievements of living with earthly variability—as it is engrained in many different cultures and ways of life. For all of us, in our own ways, are living beings whose very existence bears witness to the ability of a long line of ancestors to endure whatever the earth has thrown at them over the long march of human history and prehistory (Gunaratnam and Clark 2012).

Such an approach opens up possibilities of thinking about justice in ways that hinge not only on measurable gains and losses but on gifts or inheritances that are resoundingly incalculable. We dwell in landscapes whose rough edges have been smoothed by past inhabitants, we inherit durable material and symbolic cultures. And we are all living beings whose bodily capacities come to us "weathered" and "worlded" (Niemanis and Loewen Walker 2014) through the chains of bodies that precede us. In these ways, we are all recipients of "the gift of possibility of a common world" (Diprose 2002: 141), every one of us owing a vast, immeasurable, and irrecompensable debt to all those predecessors who have made our lives possible (Clark 2010).

But such gifts come down to us deeply inscribed with inequality, the offering of some properly acknowledged while the graft and sacrifice of others is overlooked, undervalued, or just plain appropriated (Diprose 2002: 9). So once again we find ourselves drawn into a world of relations that exceed calculation, only to find ourselves obliged to do a searching and exacting accounting. For even if we are—every one of us—in debt from the very beginning, some of us are more in debt than others. Or in the words of philosopher Alphonso Lingis: "To be responsible is always to have to answer for a situation that was in place before I came on the scene" (1998: xx).

What might it mean, then, to confront climate change in terms of a responsibility not only for what I have done—or whatever actions can be pinned on me—but for what or who I am? While not directly related to climate, Peter van Wyck's (2010) account of the Dene people of Canada's Great Bear Lake region offers an example of such a responsibility. The Dene's own storying of the land, recounts van Wyck in a phrase borrowed from Walter Benjamin, moves "in rhythms comparable to those of the change that has come over the earth's surface in the course of thousands of centuries" (2010: 178). The tribe, who had an ancient intuition that something dangerous lay beneath their soil, were inadvertently drawn into the nuclear age when uranium mined from their tribal lands was used in the atomic bombs detonated over Hiroshima and Nagasaki. In spite of the Dene's ignorance of the wartime use of "their" uranium, the tribe assumed a certain responsibility for these events—and eventually elected to send a delegation to Japan to apologize for their implication in the first aggressive nuclear detonation (2010: 45).

We refer to this as exorbitant responsibility not only because it breaks out of the closed circuit of calculable exchange, but also because it responds to an earth that no longer seems to spin in predictable orbits—a planet whose very multiplicity and changeability breaks with earlier ideas of a unified "whole earth" (Clark 2016). With its demands to attend both to the rumblings of a dynamic earth and to the complex temporalities of inheritance—cultural, corporeal, ecological—it seems to us, the exploring or tracing of exorbitant responsibility potentially digs deep into the disciplinary repertoires of geographers and fellow spatio-temporal thinkers.

Exorbitant responsibility is endlessly demanding. It calls for constant attentiveness to the appeals of others and to the inevitable inadequacies of all acts of assistance. It requires calculations of which biologist-turned-climate-change-commentator Tim Flannery has observed: "(n)ever in the history of humanity has there been a cost–benefit analysis that demands greater scrutiny" (2005: 170). It takes off from a sense of unrepayable indebtedness that stretches back through an untraceable lineage of bodies into turbulent earth history.

But in the process of reaching into the receding depths of bodies, cultures, past climates, and previous phases of the earth system, exorbitant responsibility also offers drama, enchantment, inspiration. It dreams of opening the cold hard world of climate negotiation into an earth/human adventure story of unfathomable intrigue, while also recognizing that the time and energy to mobilize against injustice and to care are themselves unevenly distributed political resources (Oka 2016: 54). Excessive forms of climate justice and care, we have been suggesting, set out from a commonplace, everyday reluctance to see the suffering of others go unattended. But for many peoples, in many places, that very sense of having something to offer others is quite mundanely linked to offerings of communities past and present, of ancestors both human and more than human, of an earth enlivened by many lifeforms. In other words, it arises out of "a bond between my present and what came to pass before it" (Lingis 1998: xx), in ways that are, for many, the very stuff of daily existence.

10 | INDIGENEITY AND CLIMATE JUSTICE IN NORTHERN SWEDEN

Seema Arora-Jonsson

Theorists have argued that environmental justice requires more than just the fair distribution of environmental benefits and harms. Beyond distribution, it requires participation in environmental decisions of those affected by them. These dimensions of justice are most clearly articulated in relation to Indigenous struggles where past devaluation of place-based cultural identities is seen as a source of injustice (Hermann 2015). Climate justice is no different. The politics of culture is at the heart of climate justice. It is also clear that society's response to every dimension of global climate change is mediated by culture—meaning-making that creates collective outlooks and behavior (Adger et al. 2012; see also Chapter 8, this volume). The lack of Indigenous participation in environmental issues, despite their vital role in shaping their environments and nature–culture relations, has been increasingly acknowledged in recent decades. There is less discussion, however, on how relations with nature and corresponding differences within communities in control over land and water may become a fault line in Indigenous struggles for cultural revival. Similarly, there is little discussion on how Indigenous culture, when viewed as dynamic and imbued with politics, has little place in the everyday working of western political systems. What does this mean for climate justice and how is culture central to it?

In this chapter, I unravel some of the strands within the knotty questions of climate justice and Indigenous culture within the struggles of Sámi reindeer herders in Sweden for voice in relation to environmental issues that concern them. These struggles, and their consequent relationship with land and water, are nested within wider Sámi struggles—that include Sámi without reindeer—to reclaim a heritage that has been eroded by over a hundred years of colonization by the Swedish state. I turn to academic research as well as government and media reports on past controversies in relation to Sámi environmental rights, and interviews carried out in northern Sweden in 2008 and 2017.[1] I explore how both resource exploitation by the state and attempts at nature conservation and climate mitigation have aroused concern within Sámi communities. On the one hand, Sámi groups in their resistance to widespread mining attempts in their territories have wanted to keep fossil fuels and metals in the ground—in line with environmentalist concerns. On the other, resistance to alternative climate interventions such as wind power farms that disrupt reindeer herding have challenged governmental political priorities and policy responses to climate change. Climate and environmental justice is often associated with grassroots environmentalism (Faber 1998). However, it is also clear, and especially so here, that grassroots environmentalism is a highly contested terrain, creating new struggles over cultural belonging and control of territory. What follows is a sketch of these contestations on the ground in Sweden.

Climate change in the north and the "colonization of attachment"

For the Sámi reindeer herders, whose livelihoods and culture have depended on access to land and water, climate change portends severe consequences (Brännlund and Axelsson 2011; Moen 2008; Pape and Löffler 2012). Grazing by reindeer constitutes a natural component of Arctic-alpine ecosystems in Fennoscandia (the peninsula of the Nordic region), and ecosystem development has been shaped by grazing and associated trampling since the last glacial epoch nearly 12,000 years ago. Reindeer husbandry depends on the diversity of accessible natural pastures covering the seasonally different needs of the reindeer (Pape and Löffler 2012: 421).

Researchers warn that warmer and wetter weather may result in increased probabilities of ice-crust formations, which strongly decrease forage availability and is likely to reduce lichen availability. Thinner ice with sudden thaws and later freezes make traditional reindeer herding over water bodies more difficult. Adapting to these includes maintaining a choice of grazing sites in both summer and winter. Reindeer husbandry and the existence of many Sámi communities are contingent on the flexible use of pasture area. However, this capacity may already be severely limited because of other forms of land use (Moen 2008; Pape and Löffler 2012).

Scholars critical of the Swedish government for its failure to address the challenges to reindeer herding argue that the history of state appropriation of land for other activities has exacerbated difficulties faced by reindeer herders today (Elenius, Allard, and Sandström 2017). Brännlund and Axelsson (2011) write that the greatest constraints on the flexible strategy today, as during the colonization of the north in the 19th century, are Sámi loss of authority over land and the regulations imposed on reindeer management. Beginning in the 17th century, the area spanning parts of current day Sweden, Norway, Finland, and Russia that the Sámi called Sápmi was colonized by non-Sámi moving into the area. By the turn of the 20th century, the Sámi had become a minority. In tandem with the demographic changes, the Sámi suffered loss of authority over land and water as a result of nature protection, and the damming of rivers that submerged land.

According to Mörkenstam (2002b), in Sweden, the state has sought to define Sámi identity and their relations to land through its legal discourse since the 1800s. Legal discourses created categories such as forest Sámi, mountain Sámi, or reindeer-herding Sámi. This has had lasting repercussions on intra-Sámi and gender relations as well as on Sámi relations with mainstream Swedish society. In the modern era of what is referred to as rationalization, reindeer herding is regarded as an economic activity, leading to the establishment of Reindeer Herding Communities (RHC) as an economic unit. The recognition of only Sámi reindeer herders as rights holders leaves the majority of the Sámi people without special rights, thereby splintering the Sámi community (Mörkenstam 2002a). Divisions between the forest and mountain Sámi and reindeer herders continue to resurface over conflicts over land and water. More recently EU and international discourses on indigeneity have played a role in creating a more inclusive sense of Sámihood based on language, cultural heritage, and traditions. Legislation, however, has not kept pace with this.

With the establishment of the first national park in 1909, Sámi rights to pasture for reindeer and to hunt and fish in national parks were outlined and have continued as such into the present, despite changes in conservation policies and the management of parks (Elenius et al. 2017: 31). However, while the Sámi have usufruct rights as described above, there is almost no mention of Sámi rights in environmental codes or legislation created by the government, and no mention of Sámi rights for cultural land uses in general (Allard 2017). The lack of guarantee of Indigenous rights especially in environmental laws has left government authorities with huge discretionary powers. Over the last three decades, Sámi reindeer herders have been accused of destroying pristine nature and mountain areas, notably through their use of modern vehicles and equipment and for having herds that are too large for sustainable grazing. The environmental movement has pressed for sustainable use of reindeer grazing areas and the conservation of large predators, and have been questioning Sámi traditions within protected areas (Allard 2017: 11).

Thus, while the Sámi have been recognized as Indigenous groups, law-making does not necessarily acknowledge their presence. State policies disrupt the lived strategies of the Sámi to deal with an uncertain future. In national legislation, reindeer herding has often had to yield to other land use interests and exploitation such as dams, mining, or the making of roads that disrupt reindeer herding. Several public inquiries have pointed to the need for increased legal protection of Sámi reindeer herding (Torp 2014). According to Groves (2015: 856), the source of injustice in such disruptions should be viewed as the "colonisation of attachment." Groves (2015) draws on this concept to demonstrate how people's attachments include relations to place that extend across the social field and into the biophysical world. Planning and decision-making over land not only have biophysical implications but also affect the identity and capacity of communities to shape their futures. Place attachment may be seen as a capability that enables identity and agency. The loss of that capability would thus lead to a loss of identity and agency (Groves 2015) and the loss of lived futures and the disruptions of identity and agency that go with it.

Sámi groups have complained that the loss of authority over land cannot be compensated economically as cultural attachments go beyond the economic (Hellmark 2016; Persson and Öhman 2014). In the vignettes ahead, I show how this colonization has caused fissures in Sámi communities and also how this colonization has, in turn, been disrupted as communities have struggled for climate justice and articulated attachment to place in their struggles for rights over land.

A new climate for the Sámi? Mining and wind power

A Sámi movement has now emerged in response to political and cultural repression where the Sámi see themselves as belonging to a wider global Indigenous community and work towards achieving political and cultural self-determination within and across the Nordic peninsula. Specific cultural traditions believed to be uniquely Sámi have become symbols in an emerging pan-Sámi consciousness through renewed artistic activity in which Sámi music has been central. Through a revival of their music and art, the Sámi propose an Indigenous nature-based cosmology. This cosmology functions alongside a critique of ideas of the uninhabited north as empty spaces and an outline of the imminent dangers to Sámi way of life by climate change (Hilder 2015).

Sámi livelihoods such as reindeer grazing, hunting, and fishing have long been part of the environment in the north. Reindeer herding among the Sámi is anchored in movement, especially for pasture land from winter to summer. Over the years, herders have pointed to the increasing fragmentation of their pasture lands and disruptions in reindeer herding due to the various inroads into what the states based in the south see as uninhabited and/or unused land. I use examples of mining and wind farms to show how both resource exploitation and protection seem to continue within this frame of mind and have similar consequences. I also argue that the path that grassroots environmentalism takes needs much more critical and strategic thinking.

Resource exploitation: mining

The state granted concessions in 2013 to a Swedish company (with a multinational investor) to mine nickel in Storuman municipality. Although some rural groups welcomed opportunities that these may bring, there was a great deal of resistance to the initiative. Evoking conventions on Indigenous rights, Sámi groups in Storuman claimed the land should be reserved for reindeer herding on the basis of historical and customary rights. While both reindeer herding and mining are considered "national interests," the government decided that mining interests would take precedence over reindeer herding in Rönnbäck. Local activists wanted concessions withdrawn from the company and turned to the UN Committee on the Elimination of Racial Discrimination (CERD). In November 2014, the CERD called on the Swedish government to stop all activities at the mine pending a UN investigation on whether mining plans violate the UN racial discrimination rules (Human Rights Council 2015). In their report of 2017, they write that the opencast mining concessions granted to the company were a violation of Sámi property rights (Swedish Radio 2017). The Swedish government has rejected this decision stating that since the Sámi do not have property rights but user rights to the land, this judgment does not hold. As of the time of writing, the Sámi appeal was still pending in the Swedish court (Swedish Radio 2017). The mining project continues at a standstill but is not abandoned.

The proposed mining and the resistance to it by several groups also exacerbated longstanding differences between those who were reindeer herders (and were consulted in the process) and other Sámi who were affected by virtue of living in close proximity to the mines but were not included in the consultations. The forced relocation of Sámi families from the north to this area by the Swedish state in the late 1800s caused conflicts at the time with the local Sámi and at times these have surfaced again. Some see a continuing policy of divide and rule on the part of the Swedish state. Maria Persson, the force behind the resistance to the Rönnbäck mine and a forest Sámi asks, "What happens to the Sámi people when the government starts to put price tags on different Sámi groups in different areas of Sápmi?" (Persson and Öhman 2014: 117).

Mitigation: wind farms

Wind farms in the north have increasingly become a point of contention and conflict between local inhabitants on the one hand and energy companies and the state on the other. Although wind power had been discussed since the 1990s, a decision in 2009 in the Swedish parliament to ensure that renewable energy accounted for all energy use in Sweden by 2020 accelerated their use in Sweden.

Many of the wind farms will be constructed in northern Sweden since these areas are sparsely populated. This has invited the ire of people living in the north and especially among many Sámi who fear the consequence of yet another attempt to restrict their access to land. Once again, the wind farms cordon off land that could inhibit reindeer movement. In a report commissioned by the Swedish Environmental Protection Agency, two scientists write that it is not necessarily only the turbines that disrupt normal grazing routes but, also, the infrastructure needed around the farms because these cause barriers to reindeer movement (Skarin et al. 2013).

In a report investigating people's relations to wind power in a northern municipality, Johansson (2010) noted that the process of participation in decision-making on wind farms was far from satisfactory. Although local groups were involved in the planning process, the wind farms projects were presented more as a "fait accompli" (*fullbordat faktum*) rather than as open to debate. The report supported the Sámi claim that almost 26 percent of their winter pasture would be destroyed by the wind farms, leading to a loss of about a fourth of their reindeer. This was totally disregarded by the Ministry of Environment. The Ministry's decision stated that there was not enough knowledge about disruption caused by the wind farms and that such disruptions were difficult to predict. Thus, they saw little reason to deny permission for wind farms that they considered a national interest. The company was meant to compensate for incursions into Sámi territory (Johansson 2010). Yet again, there were complaints that only some Sámi groups were consulted in these plans while others were left out of these discussions with authorities.

According to one Sámi reindeer herder,[2] just because these lands in the north were often uninhabited, the state assumed that it was fully permissible to start building on them. He pointed out that these lands were indispensable for reindeer movement. In an interview in May 2017, another reindeer herder regretted that government authorities had difficulties in understanding reindeer herding, especially in grasping the multiplicities of spaces that were central to Sámi herding. Besides, as he pointed out, the implications of infrastructural projects extended beyond the present and into other spaces—as reindeer moved away from areas with wind farms and made their way into other land belonging to other communities that were not within the purview of the agreements made with the company.

The question of national interest came up once again in 2016 in a case when the land and environmental court in Sweden ruled in favor of the Vilhelmina reindeer herding community. The ruling acknowledged that wind farms in Åsele in the north would disrupt reindeer herding in the area and that the company would not be able to provide for protection measures that guarantee the safety of reindeer herding. The court stated that reindeer herding is the most suitable way of promoting long-term land management and that, in this case, the two national interests—that of wind power and reindeer herding—could not be reconciled (Asmundsson 2016).

As these instances both of mining and wind farms indicate, there are no easy compromises over land such as the Swedish state hopes to achieve. Struggles over land are struggles over the meanings, identities, and the existence of a way of life. Projects of resource exploitation and climate mitigation threaten Sámi claims to land and their possibilities to imagine their future as Sámi. Climate and environmental justice are empty concepts without an acknowledgment of the histories and the relationships embedded in attachments between peoples and with the non-human world.

What is climate justice in the north? Culture/nature and control over land

It is evident that not just climate change but also climate response brings up important issues that we need to confront in working with climate justice. Climate mitigation projects that have the potential to save reindeer herding in the long run might actually undermine reindeer herding before it has a chance to be saved. For this reason, it is clear that environmentalism may be seen as a form of imperialism (Hilder 2015).

This is evident in the history of nature conservation that has divested the Sámi of their land over the years (Elenius et al. 2017), as well as through the recent intrusion of wind farms. The Swedish experience highlights how questions of "resource extraction for development" versus "land preservation for carbon sequestration" are salient priorities in the heart of a global North welfare state, otherwise known for its strong legal structures and citizen rights.

Overtures by government and company authorities to certain economic groups as representatives of Sámi culture reveal another fault line—that of nature–culture relations as imbued with politics (Arora-Jonsson 2016). This is in clear distinction to a political system where culture remains a static characteristic of a group, solely based on its economic activity. This disjuncture between culture and political representation is at the heart of many conflicts. Swedish environmental policy has had little place for culture in its work in domestic contexts in contrast to its work in the Global South through its development policies (Arora-Jonsson 2018). Even when traditional ecological knowledge (TEK) and the importance of bridging the nature–culture divide is acknowledged, everyday environmental practice has continued to entail a separation of what are considered to be natural/biophysical activities from cultural ones. Within mainstream media, Sámi "political struggles are underplayed and their role as cultural beings and TEK seems to undermine or at least kept separate from political demands" (Roosvall and Tegelberg 2013: 407). This has led Roosvall and Tegelberg (2013) to ask if this separation of politics from culture might, in fact, be one of the major obstacles for serious representation at climate summits?

According to some, the favorable focus on Indigenous knowledge has made it incumbent on groups such as the Sámi to use language—of Indigenous people as victims, as close to nature, and as carers of the environment, with its colonial echoes—to legitimize their involvement in the management of local resources, and, as a means to emancipate themselves (Maraud and Guyot 2016). However, one could also argue, as does Hermann, that it is more appropriate that the Sámi's use of language takes over colonial discourses and constructs a new, temporally layered understanding of Sámi life. The displacement of colonial narratives of victims and viewing Indigenous peoples as closer to nature, and their replacement with the Indigenous narratives of custodians of the north, will allow policy makers to see the north as a place stripped of its uncivilized and empty characteristics, and one that needs protection from the state (Hermann 2015: 377–99).

However, these discourses also reveal inherent contradictions related to control over territory within Indigenous groups—between those with rights over land and those without—who are embroiled in a political struggle for a Sámi identity. Contesting mainstream official rhetoric that there is land for all, the chair of the Sámi parliament pointed out that there is not in fact enough space for everyone. Stating that the Sámi way of life has had to give way as a result of 100 years of colonial aggression, he appealed to the Sámi to promote Sámi rights in an uncompromising way, realizing that it would imply a loss of income for those who have adapted to other ways, such as working for mining companies. He notes that many Sámi have become dependent on this unsustainable monetary economy, and that it is now an addiction that is difficult to cure (Mikaelsson 2014: 85). How these relationships will take shape in different contexts remains to be seen, but these are nonetheless central to how we think and work with climate justice.

Thus, Adger and colleagues' (2012: 115) observation that locally-based decision-making incorporates culture by building on local social norms and effecting change from within is somewhat limited. Climate justice cannot be assumed for a culture, without studying the political struggles that make that culture. Further, no culture is isolated. Places and cultures are connected in different ways and degrees to what we may call global currents (Arora-Jonsson 2016), as Indigenous movements have demonstrated around the world. Work on climate change needs to contend with new forms of grassroots political organizing that are local but also go much beyond the local sites, and are also open to social norms across scale. Creating narratives also means creating space for agency. Different Sámi groups have been part of creating a new sphere where they demonstrate that the Sámi also hold expertise and need to be heard. These challenge static ways to confront climate change, for example in the case of wind power. Conventional narratives on mitigating climate change could obscure these other discourses.

A multivocal narrative is at the heart of climate justice in the north and people's movements that are constantly redefining the narratives need to be understood. This entails a need to aspire for climate futures (in the plural) and pay attention to unequal power relations that obscure contending narratives. Climate injustice relates to this loss of "lived futures" (Groves 2015) and the disruptions of identity and agency that go with it. There is thus an urgent need to reimagine the space for the political and the new "politics of belonging" (Yuval-Davis 2011). Place attachments and subjectivities are at the heart of such a politics. They are also at the heart of the politics of culture and are essential as part and parcel of dealing with climate justice.

Notes

1. Projects on Sustainable Development and Environment. The interviews referred to here were carried out during the course of two projects between 2008 and 2015 and 2016–2017 and financed by the Swedish Research Council, Formas, and Vetenskapsrådet. The interviews were carried out by the author with the exception of four interviews in 2017 with reindeer herders in Arvidsjaur by Kennie Sandström Stridsby.

2. Statement made at a seminar on reindeer herding, Hugo Valentin Centre, Uppsala University, January 24, 2017.

11 | OUT OF THE CLOSETS AND INTO THE CLIMATE! QUEER FEMINIST CLIMATE JUSTICE

Greta Gaard

Dancing to salsa and merengue at Pulse nightclub on Latinx night in June 2016, the queer and trans* community of Orlando, Florida was unprepared for the gunman who arrived at 2:00 a.m. with an assault rifle and a pistol, killing 49 people and wounding 53 more. The violence continued from 2:00 to 5:00 a.m., with the 29-year-old shooter claiming his actions honored the Islamic State of Iraq and the Levant (ISIL). While mainstream media heralded this as the deadliest mass shooting in US history, even grieving activists were quick to cite the Battle at Wounded Knee in 1890, where US cavalry massacred 300 Lakota Sioux—mostly unarmed women and children—and to recall the racially motivated shooting just a year earlier at the oldest Black church in Charleston, South Carolina, where nine members of Bible Study group (half of whom were seniors) were murdered by a 21-year-old white man who claimed he wanted to start a "race war."

In the days that followed, the intersections of race, sexuality, and environment were eloquently articulated in e-mails from environmental and climate activists. Michael Brune (2016) of the Sierra Club expressed sorrow and solidarity, affirming that "standing boldly against homophobia, transphobia, racism, Islamophobia and sexism is the only way we can tear down the systems of oppression and exclusion that have divided our country for far too long." The internationally known climate justice organization, 350.org (Capato et al. 2016) sent out a collectively authored message of grief and hope, affirming "our fights are connected," and "as LGBTQ+ climate activists, we need to bring our whole selves to this work." Disclosing that "many of us who are shoulder to shoulder with you in the streets are LGBTQ+," 11 queer and trans* activists of color from 350.org provided the climate justice movement with its first nationally publicized coming-out statement.

Against nature?

Work connecting the feminist, environmental justice, and climate justice movements had been ongoing for decades prior to the tragedy at Pulse nightclub. Ecofeminist writings in the 1990s had initially addressed the colonialist links between homophobia and ecophobia, Western culture's fear of nature and all those socially constructed as "closer to nature," whether by gender, race, or species. As ecofeminists argued, colonialism provides a clear example of the intersections and devaluations of all those associated with nature, whether by indigeneity, race, sexuality, gender, or species.

So, why have arguments against homosexuality have always involved appeals *to* nature? Queer theorists who explore the natural/unnatural dichotomy find that "natural" is invariably associated with "procreation," an equation all too familiar to feminists. Refusing childbearing through alternate sexualities or birth control (including abortion), women are described as "unnatural," and queer sexuality is seen as "against nature." Such arguments imply that nature is valued, yet Western culture has constructed nature as a force to be dominated if culture is to prevail. These contradictory claims reveal that the "nature" queers are urged to comply with is simply the dominant paradigm of heterosexuality.

In fact, the ample evidence of same-sex sexual behaviors in other species confirms that such behaviors transcend the procreative. Popular science books such as Bruce Bagemihl's *Biological Exuberance* (1999) and Joan Roughgarden's *Evolution's Rainbow* (2013) document a vast range of same-sex acts, same-sex childrearing pairs, intersex animals, and multiple "genders." Female homosexual behavior has been found in chickens, turkeys, chameleons, and cows, while male homosexual behavior has been observed in fruit flies, bulls, dolphins, porpoises, and apes. Like sexuality, mating behavior varies across mammal species: some pairs mate for life (jackals), some have multiple partners (zebras, whales, chimpanzees), and some are homosocial, seeking out members of their species solely for procreation. The protests against the New York City Zoo's male chinstrap penguin couple who hatched a penguin egg and raised the offspring as their own (see Smith 2004) show both the depth of homophobic fears about gay parenting, and the wisdom of inspired zookeepers around the world, who have subsequently allowed same-sex penguin couples to adopt eggs.

To destabilize heteronormativity even further, queer approaches to plant studies reveal that plant species display a range of behaviors in reproduction, kinship, and association that rival that of animals: triads, multiple partners, self-pollination, and multiple genders all exceed compulsory heterosexuality's mandates in their queer botanical vitality. Because so many species have their own sexualities and cultures that don't fit with dominant human cultural models, it appears impossible to require humans to comply with "nature," for which species' "nature" would be the model? Would it be the black widow spider, who eats the male after mating, or the praying mantis, who eats the male while mating? Would it be the lesbian lizards, who reproduce by virgin birth? Evidently, attempts to naturalize one form of sexuality above all others are, at root, attempts to foreclose investigation of sexual diversity and sexual practices. Such attempts manifest Western culture's homophobia, erotophobia, and ecophobia.

Climate change homophobia

Climate change homophobia is evident in the media blackout of LGBTQ+ people in the wake of Hurricane Katrina in 2005, an unprecedented storm and infrastructure collapse which occurred just days before the annual queer festival in New Orleans, "Southern Decadence," a celebration that drew 125,000 revelers, when conservative groups first began videotaping the event and filing a petition to have it terminated. In 2005, the religious right quickly declared Hurricane Katrina an example of God's wrath against homosexuals, waving signs with "Thank God for Katrina" and publishing detailed connections between the sin of homosexuality and the destruction of New Orleans (Richards 2010).

Queer and transgendered persons already live on the margins of most societies, often denied rights of marriage and family life, denied health care coverage for partners and their children, denied fair housing and employment rights, immigration rights, and more. Climate change exacerbates pressures on marginalized people first, with economic and cultural elites best able to mitigate and postpone impacts. As a global phenomenon, homophobia infiltrates climate change discourse, distorting our analysis of climate change causes and climate justice solutions, and placing a wedge between international activists. For example, at the First World People's Conference on Climate Change and Mother Earth held in Cochabamba, April 19–22, 2010, Bolivian President Evo Morales claimed that the presence of homosexual men around the world was a consequence of eating genetically modified chicken: "The chicken that we eat is chock-full of feminine hormones. So, when men eat these chickens, they deviate from themselves as men" (Valenza 2010). This statement exemplifies a nexus of sexism, speciesism, and homophobia that overlooks the workings of industrial agribusiness, and simultaneously vilifies gay and transgendered persons as unnatural "genetic deviants." And it illustrates the need for queer feminist climate justice—because all our climates are raced, gendered, and sexualized, simultaneously material, cultural, and ecological.

Given the correlation and mutual reinforcement of sexism and homophobia (Pharr 1988), it should be no surprise that the standpoints on climate change for women and LGBTQ+ populations are comparable. While skeptics have debated whether a higher participation of women leads to better climate policy, and whether there is any verifiable gender difference in climate change knowledge and concern, the data suggest that women would make different decisions about climate change problems and solutions (see Table 11.1).[1]

While gender balance at all levels of climate change decision-making is necessary, it does not automatically guarantee gender-responsive climate policy. A wider transformation is needed, involving progressive men and genderqueer others.

Very few studies have recognized a *queer* ecological perspective, much less brought that perspective to climate change research and data collection. Yet according to a US poll conducted by Harris Interactive, "LGBT Americans think, act, vote more green than others" (2009).[2]

Table 11.1   Gender Differences in Climate Change Knowledge, Attitudes, and Actions

- Women are estimated to compose between 60 percent and 80 percent of grassroots environmental organization membership, and are more active in environmental reform projects.
- Women tend to perceive environmental risks as more threatening and express greater concern about climate change than do men.
- Women in the US show greater scientific knowledge of climate change, approach the issue of climate change differently, and express different concerns and potential solutions to problems.
- Women consider climate change impacts to be more severe.
- Women are more skeptical about the effectiveness of current climate change policies in solving the problem, whereas men tend to put their trust in scientific and technical solutions.
- Women are more willing to change to a more climate-friendly lifestyle.
- Climate protection policy areas—energy policy, transportation planning, urban planning—tend to be male-dominated.
- Women are underrepresented in areas of climate change policy.
- Women underestimate their climate change knowledge more than do men.

Most significant in the Harris Poll—given that heterosexuals are more likely to have children—was the LGBT response expressed for what kind of planet we are leaving for future generations, a question which concerned LGBT respondents at 51 percent as compared with 42 percent of heterosexual respondents. Yet in United Nations discourse to date, when LGBTQ+ people seek an entry point into the ongoing climate change conversations, the primary entry point is one of illness, addressing only HIV and AIDS. From these studies, it appears that structural gender inequality, and more specifically the underrepresentation of women and genderqueers in decision-making bodies on climate change, is actually *inhibiting* national and global action in addressing climate change. For example, research has shown that women and queers have more environmental consciousness than straight white men—yet women and queers are largely excluded from decision-making bodies addressing climate change. Their exclusion means that the decisions affecting all of us are lacking both information and analysis that do not exist separately from the embodiment of diverse communities (see Table 11.1).

The culturally constructed fear, denial, and devaluation of our embodied erotic is not lost on eco-activist youth, who are among the first to mention sexual well-being in climate change discussions. At COP 18 in Doha, Qatar, November 26–December 8, 2012, a Youth Gender Working Group emerged, emphasizing issues like the right to financing and technology, and how disasters impact women, LGBTQ+ communities, sexual health, and reproductive rights. Updating the Gender and Climate Change Network's slogan, youth agreed, "there will be no climate justice without *queer* gender justice" (De Cicco 2013).

Queering climate justice

The intersections of gender, sexuality, race, and economic justice have not always been a recognized part of either feminism or the environmental justice movement, but the internal diversity of these movements has been present from the start. LGBTQ+ people have taken leading roles in many movements, from the women's suffrage and abolitionist movements of the 19th century, to the Harlem Renaissance of the 1920s and 1930s, to the Civil Rights movement, and the social movements of the 1960s. But in these movements, queer activists have not been recognized for their work as LGBTQ+ people, and have had to form separate organizations addressing their multiple needs and interests.

In 2002, an international coalition of activists drafted the *Bali Principles of Climate Justice* (2002), articulating the links between climate change and environmental justice, and expanding the 17 *Principles of Environmental Justice* (1991) to a list of 27 principles, specifically including the rights of women, youth, and the problem of ecological debt. Another formal expansion of climate justice that includes LGBTQ+ rights has yet to be developed, but queer climate justice organizing has already begun.

At the 2014 Fossil Fuel Divestment Convergence at San Francisco State University, a workshop on "Queering the Climate Movement" was offered to participants who wanted to explore the question, "how come our movement so rarely talks about the intersection of queer, trans* and climate justice?" Participants concluded that not only are there groups within the climate movement that have not recognized the intersections between LGBTQ+ and climate movements, but that queer and trans* folks also need to be more visible, both in the climate movement and in allied movements defending justice for marginalized communities. Among those allied movements should be #BlackLivesMatter and #IdleNoMore, for as Naomi Klein (2014b) observes, "the reality of an economic order built on white supremacy is the whispered subtext of our entire response to the climate crisis, and it badly needs to be dragged into the light." Such alliances will educate queers, feminists, and climate activists about the depth and complexity of racism, which is still invisible to those in dominant groups—and many in marginalized groups as well.

At the People's Climate March in September 2014, a workshop on "Queers for Climate" was offered to articulate the ways that queers are affected by climate change, the reasons queers need to act on climate change, and how queer activism has tools that can be useful to the fight for climate justice. Discussion addressed the fact that in crises, already marginalized groups experience added stress: women face increased levels of sexual assault, while queers and people of color face higher levels of discrimination, prejudice, and hate crimes. "What good is marriage equality on the Titanic of climate change?" participants asked. And queers can't build racially inclusive communities on the ground of toxic and stolen lands. There are many skills queers bring to the climate crisis, developed over a history of mobilizing for human rights and for federal responses to the AIDS health care crisis, along with the multitasking skills of addressing government institutions, cultural homophobia, and public opinion while continuing to do the everyday work of maintaining community—providing companionship to elderly, sick, or disabled queers along with problem-solving support, food, transportation, and laundry.

Despite these skills, at the 2014 People's Climate March, queer and trans* people of color (QTPOC) were still backgrounded. As an organizer with the Audre Lorde Project, Ceci Pineda explains that the lineup for the People's Climate March placed LGBTQ+ folks in the seventh and final group, far behind communities identified as on the "frontlines of the crisis and forefront of change," or those who "can build the future," "have solutions," and "know who is responsible." This placement erases the knowledge and experiences of queer and trans* people of color and reopens wounds from the larger homophobic and racist culture as well as the LGBTQ+ culture. The Audre Lorde Project had just completed a solidarity letter naming the harsh impacts that climate change has on the queer and trans* people of color community, as marginalized communities bear increased burdens at all stages of environmental disasters, and are particularly vulnerable to climate change's adverse health effects, and its negative impact on food supply and drinking water. Low-income queer and trans* folks, particularly those with disabilities or mental illness, are even more vulnerable. Queer and trans* politics recognize that confronting climate change requires challenging the systems of oppression that exploit the earth and most human communities.

Influenced by the larger cultural forces that queers both critique and resist, queer organizations may still reflect the segregation of the larger US culture. In recognition of this problem, many organizations working on racial equity also engage in lesbian, gay, bisexual, and transgender (LGBT) advocacy, but these examples are not highly visible and are often under-resourced, a point made in the 2013 Applied Research Center *Race Forward Report* discussed below. As a result, misperceptions about the potential linkages between racial and LGBT justice flourish, including assumptions that few people of color identify as LGBT or that people of color are more homophobic than whites. In their 2013 *Better Together* report, Arquero, Sen, and Keleher of the Applied Research Center (now renamed Race Forward: The Center for Racial Justice Innovation) recommend:

- Expanding media visibility and communications capacity of LGBT people of color and those working at the intersection of race and sexuality,
- Developing LGBT leaders of color,
- Increasing support for strategic political analysis that links racial justice and LGBT equity, and
- Investing in tools that expand beyond specific policy fights like marriage equality and Don't Ask, Don't Tell to engage long-term capacity development and coalition building (Arquero, Sen, and Keleher 2013: 6).

The road to education about racial, gender, and sexual justice includes working within existing organizations to build broader pathways toward intersectional analysis and activism. Groups such as the gay and lesbian Sierra Club chapters in California, Colorado, and Washington, the US-based Queer Farmer Film Project, San Francisco's Rainbow Chard Alliance, Toronto's EcoQueers, and Minnesota's Outwoods all bridge the queer/environmentalist communities that are working to address white-hetero-privilege in the environmental movement. Given the persistent racial segregation of the United States, LGBTQ+ organizations within diverse urban centers or with a national reach are more able to organize queer and trans* communities of color. The Washington state-based Out4Sustainability, for example, has chapters in San Francisco Bay, Phoenix, Vermont, and New York, and organizes annual Earth Gay service projects, an annual Fab Planet Conference, and Greener Pride. Publisher of the daily news site ColorLines, Race Forward organizes the Facing Race National Conference and provides mobilization, skill-building, leadership development, organization- and alliance-building, issue-framing, and research reports. While the climate justice movement clearly addresses racial and gender justice, its ability to integrate an intersectional approach that foregrounds climate impacts on queer and trans* communities of color is still in progress.

Queer alternatives to climate crises

Creating sustainable and just alternatives to climate change is a crucial part of the climate justice movement, and queers are contributing in many ways. Queer food justice grows out of today's budding eco-queer movement, and is shaped by queer farmers and gardeners who feel uncomfortable in the mainstream white, heteromale, and middle-class locavore movement. The grassroots food justice movement is far from this stereotype, and reaches back to Black women rural gardeners in the post-Reconstruction South and in Harlem's rooftop gardens. In San Francisco, Queer Food For Love (QFFL) provides food, community, and a safe space against prejudice, while the Rainbow Chard Alliance bridges the organic farming movement and the queer movement, creating community for like-minded "eco-homos" in the Bay Area. In the US, the queer food justice movement is articulated through groups ranging from Vermont, Massachusetts, California, and Connecticut to Tennessee, Alabama, Arkansas, Kansas, and Washington. Concerned about the intersections between environment, sexuality, and gender, these queer groups use food to build community, fight oppression, and take care of planetary and human bodies.

From an intersectional standpoint, food justice cannot be defined solely in terms of justice across human diversities, excluding those who count as "food." Founded by an Arab-American and a white working-class lesbian couple in 2000, VINE Sanctuary is run by five queer and trans*-identified activists, providing a haven for animals who have escaped or been rescued from the meat, dairy, and egg industries, or other abusive circumstances, such as cockfights or pigeon-shoots. Sanctuary residents include chickens, cows, ducks, doves, geese, pigeons, sheep, emus, and even a few parakeets. In addition to sheltering and advocating for animals, VINE conducts research and provides education aimed at creating systemic changes in agriculture, trade, and consumption as well as human attitudes about animals and the environment, as these intersect with racial and gender justice.

Just as the exploitation of animals initially set the stage for race-based exploitation of people, VINE recognizes that today's racial and economic injustices perpetuate both environmental racism and the continued exploitation of animals. Dangerous and environmentally destructive factory farms and processing plants are often located in communities of color. Local citizens must live with the pollution while working at dangerous and degrading jobs. The products of these industries are often marketed to communities of color, regardless of the impact on physical health or cultural welfare. US dietary guidelines recommending high consumption of meat and dairy products are a form of *food racism*, as up to 95 percent of adult Asians, 74 percent of Native Americans, 70 percent of African Americans, and 53 percent of Mexican Americans are lactose intolerant (Gaard 2017: 61). Recommendations concerning meat consumption ignore the high rates of heart disease, hypertension, and diabetes among African Americans. Low-income communities of color continue to be the sites of high concentrations of fast food restaurants (i.e., "food deserts") without grocery stores offering fresh fruits and vegetables, bulk grains, and other inexpensive ingredients for a healthy diet.

Finally, queer cultural skills such as queer aesthetics, queer performativity, pageantry, drag, and polymorphous perversity are all tools useful to the climate justice movement. Ecosexuality is but one example of the activist potential for queer feminist climate justice. In their eco-documentary, *Goodbye Gauley Mountain* (2013) Beth Stephens and Annie Sprinkle expose the ways that coal mining and mountaintop removal affect queers, working-class poor people, and ecosystems. The Black miners who died of silicosis in the 1930s, working to build Hawks Nest Tunnel for the coal company; the mountaintop-removal communities of poor white people who have a 50 percent increase of cancer, and are 42 percent more likely to have children born with birth defects—both illustrate the "slow violence" of rural environmental racism and classism, wiping out the culture and ecocommunities of West Virginia, where a monoeconomy keeps people in thrall to the coal industry. Bringing a homegrown queer performance artist like Beth Stephens and her wife and former porn star, Annie Sprinkle, to West Virginia's embattled mining communities, Beth and Annie's ecosexual weddings bridge the urban/rural, queer/straight, white/people of color schisms by affirming a shared and longstanding love of the mountains, and celebrating that love in drag and polyamorous commitment. Ecosexuality "shifts the metaphor from earth as mother to earth as lover," says Annie Sprinkle, "to entice people to have more love of the planet" (Stephens and Sprinkle 2016).

For a queer and present climate justice

The tragedy at Pulse nightclub effectively outed the climate justice movement. As Suzanne Pharr explained in 1988, homophobia is rooted in cultural misogyny, and the liberation of women, people of color, and queers are inextricably interconnected. Moreover, an economic system reliant on enslaving people, animals, and the earth cannot survive. Nature is far from heteronormative, and real climate justice will have to include all of us.

Notes

1. The careful methodology of these studies affirms their validity. International findings on gendered differences in climate change causes, analyses, and solutions in Ergas and York (2012) rest on 60 peer-reviewed studies, which then shape the questions and statistical analysis these authors undertake. McCright (2010) tests the arguments about gender differences in scientific knowledge and environmental concern using eight years of Gallup data on climate change knowledge and concern in the US public. Alber and Roehr (2006) report on the project "Climate for Change—Gender Equality and Climate Policy" that performed data surveys of the gender balance in climate policy at local and national levels for ten major cities in four European countries (Germany, Italy, Finland, Sweden). The studies are cited and discussed in Gaard (2015). I recognize that "gender" here is defined in binary terms and searched for but could find no research on trans* perspectives about environmental issues, though I suspect these perspectives would be comparable to other non-dominant gendered views.

2. Because the findings may surprise some readers, I include links to Harris Interactive Methods for LGBT surveys: www.harrisinteractive.com/MethodsTools/DataCollection/SpecialtyPanelsPanelDevelopment/LGBTPanel.aspx.

PART THREE

SOCIAL SCIENCES, HUMANITIES, AND CLIMATE JUSTICE

12 | SLEEPWALKING IS A DEATH SENTENCE FOR HUMANITY: MANIFESTO FOR A SOCIOLOGY OF THE CLIMATE CRISIS AND OF CLIMATE JUSTICE

John Foran

> Many intellectuals in the social sciences and humanities do not concede that Earth scientists have anything to say that could impinge on their understanding of the world, because the "world" consists only of humans engaging with humans, with nature no more than a passive backdrop to draw on as we please. (Clive Hamilton 2017b)

New realities have always called for new paradigms, and sociology—the study of how societies are structured by inequalities and how they might change—is built on the foundational work of giants like Karl Marx and Max Weber, who grappled with explaining the rise, functioning, and possible future of capitalism as it burst onto the scene in the 19th century. The most original and critical works of 20th-century sociology did a decent job of keeping up with the great changes that followed: corporate control of the global economy, the great social revolutions and other attempts to make societies fairer and more just, the rise of social movements demanding rights for women, for people of color, for gendered others, for all humans' rights generally, and now humans' responsibilities toward animals, the planet, and the very future we hope to have.

But even while doing so, much of the discipline lost its critical punch, and nowhere has this been so dramatic and fateful as in the inattention of both mainstream and critical sociologists alike to issues of environmental and climate-induced destruction as the 21st century has rolled into being, and as their effects have inexorably become inescapable realities.

The present moment is a dangerous one, calling for our immediate attention, as the academy, like humanity itself, now must scramble to address the existential question of climate change, and with urgency. What has been left to a new generation of younger scholars and activist intellectuals is to trace the contours of the new field we need going forward: a sociology of the climate crisis, and a sociology of climate justice, alongside the movement that bears that name. In other words, the challenge for this generation and its allies is the forging of a completely new understanding within every field of sociology and across the social sciences, the environmental humanities—our close allies in this—and for that matter, all the fields of the natural sciences, engineering, and education, both in the university and among public intellectuals and movement strategists.

This is a tall order. But as radical journalist Chris Hedges (2016) has put it:

> It is up to us to resist. We must refuse to be complicit, even in the act of voting, with the fossil fuel industry's savaging of our ecosystem, endless wars, oppression of the poor, including the one in five children in this country who is hungry, the evisceration of constitutional rights and civil liberties, the cruel and inhumane system of mass incarceration and the state-sponsored execution of unarmed poor people of color in our marginal communities.

Those who respond to the crises of our times must reckon with such questions as:

- What might a sociology of climate change look like?
- How and why might the global climate justice movement become the biggest, most transformational, and consequential global social movement of the 21st century?
- Is it incumbent now to change the way we practice sociology, and if so, what does this mean and how might we do it?

In other words, what are some of the specific contributions that sociology might make to the creation of a strong new field of climate justice studies?

The facts: readily available climate science is extremely clear that the "business as usual" (BAU) global economic model of the current phase of capitalist globalization will take average world surface temperatures past a two degrees Celsius rise, this being best understood as the boundary between "extremely dangerous" and *chaotic* degrees of climate change (see Anderson 2012, and any analysis of the national pledges that underlie the 2015 UN climate summit's *Paris Agreement*).

The situation is far worse, though. Climate science has established the threshold of 1.5 degrees as a far safer upper limit for (hopefully) staving off chaos in the Earth system. And remember too that we are here only talking of average surface temperature warming: what counts is the impact of this on our food systems, fresh water supplies, polar ice, the very air we breathe, ocean life, the increase in devastating storms, droughts, floods, and fires, all of which will vary by region. Elementary social science suggests that our own and subsequent decades will witness dangerous social consequences, including massive migrations of up to several billion people away from flooded coastal cities, chronic wars and many other forms of violence over food, land, and water, and quite possibly the end of democracy in any of the forms we know today toward even more authoritarian, invasively surveillant, militarized states across most of the globe.

A lucid look at these elementary "social facts," as Durkheim (1982 [1895]) might have called them—the values, cultural norms, and social structures of a given society[1]—suggests that variants on these scenarios become increasingly probable with every tenth of a degree rise above 1.5 degrees. If these analyses are accepted, sociologists need to ask "what kind of carbon budget keeps us under 1.5?" That is, how much greenhouse gas can humanity put into the atmosphere before we pass that mark? Perhaps this account provides the most succinct overview of our predicament:

> If the world pursues the Paris Agreement's more ambitious limit of 1.5C, the timescales over which global emissions need to peak and start falling rapidly are much shorter.
>
> ...
>
> [T]here are just over *four years' worth* of current emissions left before it becomes unlikely that we'll meet the 1.5C target without overshooting and relying on unproven "negative emissions" technologies to remove large amounts of CO_2 out of the air later in the century. (Pidcock 2016, emphasis added; see also Carbon Brief 2017)

Thus, my call to the academic world (and not just that world): *it's now time to wake up and dedicate our collective lives to this civilizational crisis.* And that means pretty much dropping BAU scholarship in the BAU university. Instead, educators and students could bring about a massive transformational mobilization of education and civil society at every level, both qualitatively, in what and how we teach and work, and quantitatively, in whom higher education reaches and whom it excludes.

Now, in the age of the Anthropocene, BAU thinking will be blown away by the insistent winds of climate change, and by all indications, so might capitalism itself, and quite possibly our taken-for-granted "civilizations," or even humanity itself. And given the existence of abrupt climate change triggers—such as an ice-free Arctic releasing vast stores of methane from the ocean floor—we may be facing these conditions well before 2050, the usual date proposed for driving greenhouse gas emissions down toward zero (McPherson 2017; Wadhams 2017; Wallace-Wells 2017).

A sociology of the climate crisis

What might a sociology of climate change look like?[2] Perhaps it should start with the economic bedrock of the current situation.

> The bottom line is what matters here: our economic system and our planetary system are now at war. Or, more accurately, our economy is at war with many forms of life on earth, including human life. What the climate needs in order to avoid collapse is a contraction in humanity's use of resources; what our economic model demands to avoid collapse is unfettered expansion. Only one of these sets of rules can be changed, and it's not the laws of nature. (Klein 2014a: 21)

Neoliberal capitalism's multiple crises are mostly the effects of its *normal* operations. In the last 20 years, the rampant privatization of public goods and services has generated obscene inequality and unparalleled concentrations of wealth and power: while just 90 corporations and fossil-fuel-exporting countries are responsible for fully two-thirds of all the carbon emissions discharged since the dawn of the Industrial Revolution, so the richest five individuals in the world now possess as much wealth as the poorest half of humanity—three-and-a-half billion people (Wearden 2014; the original study is by Oxfam 2014).[3] To this we may add Rob Nixon's (2013) "slow violence" of resource depletion, violence, and militarism. With climate change now in the ascendant, we are entering the stage of the coming crisis of capitalism.

The sociology we now need may be defined as the study of everything about this crisis and its possible trajectories. And the first principle of this sociology would be that everything is connected to everything else; curiously, perhaps, this would be one way of stating the first principle of ecology, or the worldview of Buddhism (Batchelor 1997).

Here are some of its characteristics.

Multiple crises

One way to think of the present moment and the foreseeable future is as a *triple crisis* of economics, democracy, and pervasive violence. We are living in real time through several profound deteriorations in the quality of life on Earth:

- A deep and persistent economic crisis is manifest in global economic uncertainty and unequal access to well-being in the current period, including vast gender and racial/ethnic/national disparities;
- A crisis of democracy is growing as governments and political parties fail to live up to public expectations in many regions of the world; and
- Violence and militarism saturate the global economy and national cultures as a result of both the economic crisis and the democratic deficit.

The nodes of this triple crisis are bound together by and exacerbate the likelihood of climate chaos. The interdependency of the several crises means that holism is essential to confront climate change, and to advance movements for climate justice. And this implies the joining of intersectional analysis and activist scholarship.

Climate change as a "wicked" problem

As today's intertwined crises are inextricably linked and bound together, they are exacerbated by the wild card of climate chaos, auguring a perfect storm of crisis (Foran 2016b). With climate change, we are faced with a "wicked" problem with no known precedent, something that is

> difficult or impossible to solve because of incomplete, contradictory, and changing requirements that are often difficult to recognize ... Moreover, because of complex interdependencies, the effort to solve one aspect of a wicked problem may reveal or create other problems ... A problem whose solution requires a great number of people to change their mindsets and behavior is likely to be a wicked problem. (Wikipedia n.d., citing the Australian Public Service Commission 2007)

While many social problems present challenges in the above sense, climate change may be considered a *super* wicked problem, characterized by four further features: (1) time is running out; (2) those seeking to end the problem (humans, and more precisely, global elites) are also causing it; (3) it is a global collective action problem overseen by, at best, a weak central authority (as anyone who has ever witnessed a UN climate summit can attest), all of which lead to the fourth obstacle:

> Partly as a result of the above three features, super wicked problems generate a situation in which the public and decision makers, even in the face of overwhelming evidence of the risks of significant or even catastrophic impacts from inaction, make decisions that disregard this information and reflect very short time horizons. It is this very feature that has frustrated so many climate policy advocates. (Levin, Cashore, Bernstein, and Auld 2012: 128)

Transboundary scholarship: creating a sociology of the future

The interdependency of the several crises besetting us is significant; it means that we need to learn to connect the dots in confronting the climate crisis, and this in turn leads us to the necessary breaking down of all disciplinary boundaries in our paths, and to become well versed in the ecological humanities, social sciences, and physical sciences. If we are trying to focus on the state of the world in the future, say in 2025 or 2050, then sociology will have to open itself to *analyzing* the future better than it currently does. The dawning of the Anthropocene compels a new relationship to time itself, at once pressing and very short term for humanity, and long-term as the planet surpasses us into deep time (Angus 2016; Bonneuil and Fressoz 2016; Hamilton 2017a).

A sociology of climate justice and the climate justice movement

To my mind, the pioneers of the sociology of climate justice include the Durban, South Africa-based scholar activist Patrick Bond (2012), and some of the key figures in its origins in the sociology of environmental justice, including Bob Bullard (1990) and David Pellow (2014).

Climate justice perspectives recognize that the brunt of climate change and environmental injustice falls hardest on the most poor and marginal peoples. The injustice of this irreducible reality implicates not only the multiplicity and diversity of responses from nation states, organizations, movements, and individuals, but also includes the recognition of culture, justice, and dignity as central to any meaningful action.

If responding to the challenges of climate justice requires the skills to imagine different futures, then a distinctive feature of the sociology of climate justice should be envisioning scenarios of change and transformation encompassing creative endeavors by artists, writers, and performers who can help re-draw the seeming inevitability of climate chaos into societies where there is hope (Milkoreit, Martinez, and Eschrich 2016; Porritt 2013; Raskin 2016). To paraphrase Naomi Klein (2014a): "To change everything, everyone must change."

Following from this, the question the global climate justice movement confronts is: just how *do* we somehow keep 80 percent of the remaining reserves of coal, oil, and natural gas in the ground, to hopefully "contain" warming in the 1.5–2.0 degree range, with the might of the world's largest corporations and richest governments fused in a determined, indeed suicidal, pact against us?

Both the depth of the current crisis, and the central role played by the climate disruption that exponentially exacerbates it, suggest that our activism around climate change may open a window to moving beyond capitalism in our lifetime. It seems increasingly evident that only *a strong and vigorous climate justice movement on a global scale* has the capacity to create governments capable of standing up to the economic and political forces of carbon capitalism, and hold governments everywhere to their commitment to social justice.

Principles, practices, dreams, and hope: constructing vibrant political cultures of opposition and creation

It could also be argued that new movements need new theories. This is true in the sociology of revolutions, for example, where the causes, processes, and outcomes of the great 20th-century revolutions differ radically from those of movements for radical social change in the 21st (Foran 2014).

None of the revolutions of the twentieth century was made without powerful political cultures of opposition that proved capable of bringing diverse social groups to the side of movements for deep social change, as happened in the Mexican, Russian, Chinese, Cuban, Nicaraguan, and Iranian revolutions (Foran 2005). These political cultures drew on people's experiences and emotions and were expressed in complex mixtures of popular, everyday ways of articulating grievances—whether in terms of fairness, justice, dignity, or freedom—and more consciously formulated radical ideologies such as socialism, liberation theology, and anti-colonialism. The most effective revolutionary movements of history have found ways to tap into whatever political cultures emerge in their society, often through the creation of a clear common demand such as "the regime must step down" or "the foreign powers must leave." The forging of a strong and vibrant political culture of opposition is thus an accomplishment, carried through by the actions of many people, and, like revolutions themselves, such cultures are relatively rare in human history.

In the 21st century, the movements for radical social change (a term more apt for this century's great social movements than revolution) has itself changed, as activists, reformers, dreamers, and revolutionaries globally have increasingly pursued non-violent paths to a better world, intending to live and act as they would like that world to be. That is, the ends of justice are no longer held to justify the means of violence, but the means of non-violent resistance reflect and guarantee the ends that they seek. In this, they embody and illustrate the virtues of prefigurative politics and in particular horizontalist ways to realize them.

We might call these positive, alternative visions "political cultures of *creation*" (Foran 2014; Foran, Gray, and Grosse 2017). Movements become even stronger when they add a positive vision of a better world to a widely felt culture of opposition and resistance, an alternative to strive for that could improve or replace what exists (Pellow 2014). In this sense, some of the differences between old and new movements for radical social change include: the attempt to get away from the hierarchical organizations that made the great revolutions of the 20th century and move in the direction of more horizontal, deeply democratic relations among participants; the expressive power of using popular idioms, visionary narratives, and compelling stories using all manner of media; the growing use of civil disobedience and militant nonviolence; the building of coalitions as networks that include diverse outlooks; and the salience of political cultures of creation alongside political cultures of opposition and resistance.

Conclusions

In the long run, the only real systemic "solution" to the crisis is a broad yet at the same time more radical climate justice movement willing to confront the root causes of the crisis, including capitalism. What is also required is the will of nation states, among others, to be strong enough to decisively cut emissions in a just way. These movements, or convergence and confluence of many movements, have to get there in the relatively medium term, say, the next quarter century. In the short term of the coming ten to 15 years, the task is to widen and radicalize climate justice movements everywhere, preparing ourselves and a new generation for the longer anti-capitalist project of deep social transformation in the direction of an ecologically sustainable, socially just, and deeply democratic global future.

The sociology that accompanies this will be rooted in climate justice studies and pursued under the banner of scholar-activism and an engaged public sociology, or it will drift into irrelevance as we are engulfed by the failure to understand what is happening to us. The path opens before us to choose the former as we wake up from the latter.

The good news is that if we deal wisely with this problem, this crisis, then we're going to build a world that works better than the one we have right now, a world that works fairly, is more democratic, less unequal. (Bill McKibben)

Notes

1. Although climate change is perhaps better considered as a "hyperobject" in the language of Timothy Morton (2013).

2. Here I must acknowledge a pioneer in this field, the late John Urry, whose 2011 book *Climate Change and Society* opened the field so thoughtfully and comprehensively.

3. In 1999, the United Nations Development Programme reported "The net worth of the world's 200 richest people increased from $440 billion to more than $1 trillion in just four years from 1994 to 1998. The assets of the three richest people were more than the combined GNP of the 48 least developed countries" (United Nations Development Programme 1999: 36–7, quoted in Prasad 2014: 234).

13 | A ROLE FOR THE ENVIRONMENTAL HUMANITIES: DIRECTLY INTERVENING IN ANTHROPOGENIC CLIMATE CHANGE

Ken Hiltner

I am a professor of English literature. I am also a professor of environmental studies. Taken apart, neither of these facts tends to raise eyebrows. However, when I inform students, friends, and even colleagues that I am indeed a professor of both, I am often met with looks of confusion. This is hardly surprising, as many people simply assume that "environmental studies" is another name for the environmental sciences. It's not. Still, the confusion is easy to understand. When departments of environmental studies first began to appear in universities nearly 50 years ago, they were largely populated by scientists, as many still are. Moreover, when it comes to an issue like climate change, were it not for the tireless work of a range of scientists and their careful observations, studies, and predictive models, the issue would never have even been discovered. Whether you praise their work or deny their findings, scientists are generally the scholars that we think of first when the issue of climate change is invoked. Conversely, I think it safe to say that environmental humanists rarely enter many minds. Even aside from the issue of climate change, the value of the humanities has repeatedly been drawn into question in recent years, especially in relation to the STEM (Science, Technology, Engineering, and Math) fields, which often appear far more practical and useful.

If, after accepting the findings of these scientists, the next step of pondering what is to be done about climate change is taken, the sciences again often first come to mind. However, in this attention is shifted from the theoretical to the applied and the hope that scientists will provide technological solutions to our worsening global problem. Consequently, it is more than a little comforting to hear techno-wizard nonpareil Elon Musk confidently proclaim that an array of solar panels just 100 by 100 miles square (160 by 160 kilometers square) (less than one-tenth of the area of either Nevada or Arizona) could power the entire US (and that a battery covering just a single square mile in area, could store all this captured energy). What is even more comforting is the larger message contained in such a proclamation: scientists have discovered a major problem that should concern us all, but just sit tight, as they are hard at work on a solution. Some, like Musk, boast that they already have it, or at least a major part of it.

So how can professors of the environmental humanities possibly help? After all, at face value it seems altogether unlikely that specialists in literature like me can save the planet from environmental devastation. In contrast, scientists can—in very specific and concrete ways—explain many of the causes of climate change such as the rise of atmospheric CO_2 and other greenhouse gases. But are these gases in fact the cause of climate change? Many likely simply assume that they are. However, I might argue that the root causes of climate change are in fact environmentally questionable human practices such as the love of wealthy Americans of transportation (especially automobiles and planes), expansive houses, endless consumer goods, and so forth, as these demand the extraction of enormous quantities of fossil fuels that release CO_2 when burned. If we curtailed these activities, CO_2 rise and climate change would be significantly checked. Similarly, if we reduced or eliminated the ravenous meat consumption of the US (it has the dubious distinction of being a world leader here), methane emissions, another major contributor to climate change, would be dramatically cut back.

Why do so many engage in these worrisome practices? Understanding why people do what they do is the domain of the humanities and the social sciences. And once human actions are better understood, possibilities of pursuing environmentally sound practices open up. In short, the humanities have a major role to play in understanding and helping limit anthropogenic (i.e., *human*-caused) climate change.

Let's take an example near and dear to the heart of many Americans: the automobile. In 2014, the efficiency of cars hit a record high, as the average automobile in the US now travels 25 miles per gallon (10.6 kilometers per litre) of fuel. Hybrids are even better, as they can be twice as efficient. Still, the motive power here, even from (non-plugin) hybrids, comes from the burning of fossil fuels. Purely electric vehicles like the Tesla are true game changers in so far as they can be run entirely on renewable energy. They are, however, not without environmental issues, as, for example, the mining and eventual disposal of rare-earth materials used in their manufacture is problematic.

Nonetheless, at the end of the day the Tesla is in many respects a rather conventional automobile (HBR 2015). The newest model, the Model X, is an SUV which, like many of its gasoline-propelled counterparts, weighs over 5,000 pounds (2.3 metric tons). Once on the road, 75 percent of these technological marvels will, like the rest of the cars being driven in the US, carry a single person. Although such a vehicle can be run entirely on renewable energy, there is no escaping physics, as propelling this much mass down the road at 70 mph (112 kph) in order to transport a single human being is incredibly energy intensive. While there has long been the hope that cars will become more energy efficient over time, and automobile manufacturers have in fact made laudable gains in this direction in recent years, it is unlikely that a practical 5,000-pound (2.3 metric ton) (or even 2,500-pound / 1.1 metric ton) vehicle will ever surpass 75 or 100 mpg (31 or 42 kpl) of gasoline or its renewable-energy equivalent.

It is, however, currently possible to transport a person 350, 500, even 750 miles on a single gallon of gasoline (149, 212, 318 kpl). This is possible because of a variety of transportation technologies that are all over 100 years old and have been in widespread use for as long. What are these wonder technologies? Buses, subways, and trains, respectively. When compared to a 25 mpg (10.6 kpl) car with a single occupant, a bus is 14 times more efficient (hence 350 mpg / 149 kpl), subway 20 times more so (500 mpg / 212 kpl), and a passenger train an astonishing 30 times more efficient (750 mpg / 318 kpl).[1]

A few years ago a student of mine, reflecting on these numbers, succinctly observed that "what we need is not a 100 mpg (42 kpl) car, but rather for taking the bus to become cool and owning a car to be anything but." I could not agree more. Incidentally, even 750 mpg (318 kpl) can be much improved upon—and it is embarrassingly easy to do so. In parts of Manhattan, over a third of commuters walk to work. The only thing that can improve the energy efficiency of walking is cycling. New Yorkers who take buses, subways, and trains to work are 11 times more likely to take mass transit to work than the average American. As Edward Glaeser (2012), David Owen (2010), and many others have argued, cities are thus far more efficient than suburbs and rural locales. This is clearly the case with energy use and corresponding carbon footprints. For example, a person in rural Vermont uses five times as much gasoline as someone in Manhattan largely because they are far more likely to drive to places rather than walk or bike there.

Why then do so many Americans drive cars? And why do we drive so many of them? The US has less than 4 percent of the planet's population, yet a quarter of its cars. Putting them end to end, they would circle the earth 31 times. As my student realized, in the US cars are cool—really cool. Conversely, with fewer than 5 percent of Americans taking the bus to work, as opposed to 85 percent using cars to commute, buses are not very cool at all. But why are cars cool and buses not?

This is not a question for the STEM fields, but rather the social sciences and humanities, where we seek to understand just why people do what they do. And make no mistake, human behavior is often enormously difficult to understand—which is why the social sciences and humanities need to be weighing in on climate change.

I will explain further, but it is first worth noting that we rightly dubbed the effect that human beings are having on our global climate *anthropogenic* or human-caused. Science may be able to tell us *how* human beings are changing our global climate, but not *why* we are doing it. The sciences may be able to offer us more technological, efficient practices (i.e., more efficient cars), but they offer little insight into why we continue to engage in these practices. Why, for example, we love cars. That is the domain of the social sciences and humanities.

If we can understand why cars are cool and buses not, we can perhaps then even take the next step—and it is a big one—of not just studying culture but actively intervening in it by exploring how we might help engineer a culture where riding a bus or train is far more appealing than a commute in a car. If we can, the gains could greatly exceed the impact that a 100 mpg (43 kpl) car will have on climate change. And riding in a bus is 79 times safer than in a car. Riding in a train or subway safer still.

This is no easy task. Creating and manufacturing the next generation of lithium batteries (one of Elon Musk's current projects) will certainly be difficult, but no less so than understanding why human beings engage in the practices that are bringing about a global change to our climate, which is sometimes as perplexing as it is irrational.

Driving a car can, indeed, border on the irrational. Riding in a 5,000-pound (2.3 metric ton) vehicle traveling 70 mph (112 kph) just inches away from others doing the same is insane. As the World Health Organization (WHO) (2017) notes, each year over 1.25 million people are killed in traffic injuries worldwide. And that is just the tip of the iceberg: as many as 50 million more are injured. Worldwide, traffic accidents are the leading cause of death for young people over the age of 10, surpassing even malaria and AIDS. In fact, the WHO has declared traffic injuries a worldwide epidemic (Peden et al. 2004).

In addition to being incredibly dangerous, automobiles take a huge portion of family income, making them far less economical than mass transportation. The average cost of owning, insuring, maintaining, and fueling a car is more than US$9,000 a year. Since three out of four working families have two or more cars, that is at least US$18,000 a year. This is a huge chunk of the median US income of US$53,000. In contrast, an unrestricted bus pass in my city (Santa Barbara, California) currently costs just US$52 per month—i.e., just over US$600 per year, which is about 7 percent of the annual cost of owning and maintaining a single car.

How is it that cars are cool? The growth of the postwar US automobile industry and the US government's unprecedented expenditure on roads and highways, which were designed to fuel this industrial juggernaut, clearly depended on convincing the public that cars were desirable. By 1960, US automobile manufacture was not only the largest industrial segment in the nation, it had become the largest industry on the planet, dwarfing anything anywhere that had ever come before it. One in six Americans were, either directly or indirectly, employed by this industry (Mieczkowski 2005: 151). And this does not include the massive, complementary industry of road construction, made possible by the ambitious Highway Act of 1956 which authorized the creation of about 41,000 miles (65,000 kilometers) of interstate highways.

Although it may sound a little outlandish on first hearing, for decades the backbone of the US economy depended on cars being cool. So cool, in fact, that we would knowingly risk our lives and dedicate significant portions of our income to their purchase and upkeep. It is difficult to imagine how a broad swathe of the American public would go along with this lose–lose proposition. More difficult still is to imagine how it continues today in an age when we are aware of climate change. The EPA (2014: 2) has calculated that a "typical passenger vehicle emits about 4.7 metric tons of carbon dioxide per year." Note that many climate scientists advise that, if we are to mitigate the effect of climate change, our greenhouse gas (GHG) emissions should not be much more than one metric ton for each person on the earth. Now that we realize that owning and using an automobile can increase our carbon footprint to over four times what it should be, it becomes even clearer that cars are anything but cool.

How can an expert at reading literature like me help with this situation? In my humble opinion, we have much to offer, as cars are not intrinsically cool, but have only been written that way. Thousands and thousands of texts, many of them car advertisements, over many years not only convinced the American public that cars are cool, but also delivered the counter but complementary and effective message that not owning a car is very uncool.

Understanding just how this happened is not only interesting, it is crucially important, as it can offer insights into how we can disrupt this decades old project. As to how this can be done, please allow me to quote myself from an earlier text where I turned to the philosopher Martin Heidegger in order to understand how we can be oblivious to certain enormous cultural influences:

> In *Being and Time*, Heidegger argues that when it comes to the intelligibility of the backdrop of our lives, we generally lack thematic awareness of this "availableness" (*Zuhandenheit*). However, if there should be a conspicuous malfunction (*auffällig*), an obstinate temporary failure (*Aufsässigkeit*), or most serious of all, an obtrusive total breakdown (*Aufdringlichkeit*) of this availableness, we would suddenly gain awareness of the backdrop as occurrent (*Vorhandenheit*), as that which emerges at hand as a presence … Applying this approach to the world of human cultures are analyses of background cultural phenomena such as patriarchy: cultural realities that are always there, yet have historically often managed to escape our attention—unless we bring about at least their partial or temporary breakdown, thereby pulling them out of the background to make them more apparent. (Hiltner 2011: 36)

In recent decades, feminist scholars have made enormous gains in drawing attention to the patriarchal backdrop of our culture. As many of these scholars are social activists, they set out to disrupt patriarchy simply by drawing attention to its complex workings, which paradoxically are often hiding in plain sight as they are so ubiquitous they often escape our attention.

While it may seem odd to compare patriarchy with the cult of the automobile, seemingly endless power and resources were (and still are) given to maintaining both. In the case of cars, our very economy depended upon it. Now, however, the future of our species and planet depends on our disruption of our relationship to the automobile and a range of similar practices. My hope is that the environmental humanities will take up this formidable challenge.

Note

1. These calculations are derived from numbers provided by David MacKay (2009), who is a Cambridge professor and former chief scientific adviser to the UK Department of Energy and Climate Change, in his excellent 2009 book *Sustainable Energy: Without the Hot Air*.

14 | MEDIA AND CLIMATE JUSTICE: WHAT SPACE FOR ALTERNATIVE DISCOURSES?

Anabela Carvalho

Injustice and awareness

As put by one of the most prominent and vocal climate researchers, climate change is, profoundly, an ethical issue. "Today's changes of atmospheric composition," James Hansen and colleagues write, "will be felt most by today's young people and the unborn ... who currently depend on others who make decisions today that have consequences over future decades and centuries" (Hansen et al. 2011: 22). Awareness is a game-changer in determining the moral (un)acceptability of this and other forms of injustice (e.g., international, inter-class) embedded in climate change: "Our parents honestly did not know that their actions could harm future generations. We, the current generation, can only pretend that we did not know" (Hansen et al. 2011: 22).

Surveys conducted around the world show that most people are indeed aware of climate change and multiple studies suggest that this is largely due to the media. From this perspective, the media are crucial to generating shared views on climate-related injustice. Although some studies suggest that citizens are indeed concerned with issues of injustice in climate-related matters (McLaren et al. 2016), it is not known how widespread such feelings are and how much they weigh on (individual) decisions. Obviously, there are multiple factors at play but the discourses that the dominant media have co-constructed in the last few decades are a key factor for social representations. At the same time, a growing number of social movements struggling for climate justice have been using various communication tools, including "alternative media," to disseminate counter-hegemonic views.

The importance of the media for the definition of the meanings of climate change derives from the place they occupy in current-day public spheres. News media, specifically, are a crucial space for the amplification of the viewpoints and arguments advanced by multiple social actors as well as a key agent in the construction of discourses in their own right. However, a few words of caution are due on what is meant by "the media." Firstly, although many reflections on the social and political roles of media discourse speak of the media as a unitary and homogeneous body, it is necessary to differentiate between the numerous, and profoundly diverse, institutions that make up "the media." Their goals, audiences, and channels, to name just a few aspects, vary widely and the range of "alternative media" projects that have been developing in the last few decades is evidence of increasing pluralism despite a simultaneous tendency for concentration of property. The second caveat is that the vast majority of extant research is on "Western" countries, particularly in Europe and North America, whereas there are nearly no studies on the countries most vulnerable to climate change (Schäfer and Schlichting 2014). It should also be noted that over two-thirds of research has focused on print media, whereas television, radio, and internet-based media have received much less attention until very recently. The trends reported here draw on such published research and therefore only yield light on a small part of the world media landscape. Third, news and journalism, although the main focus of this chapter, are not the only modes of mediated communication that have "political" meaning and that shape understandings of the world. Instead, citizens make sense of social and political reality through a profusion of media and types of content. Films, video games and documentaries are among the formats shown to have played a role in relation to views on climate change (e.g., Lowe et al. 2006). Finally, whereas so-called "social media" have generated much enthusiasm in some circles, it has also been shown that they often operate as echo-chambers for the "mainstream media" on climate change and other issues (Kirilenko and Stepchenkova 2014), and in some cases have mined the efficacy of movement communication (Poell and van Dijck 2015).

Reinforcing the "order of things"

Dominant media(ted) discourses on climate change have mostly reinforced rather than challenged the order of political, economic, and social things. Drawing on Foucault (2002), I argue that the mainstream media have contributed to create a system of intelligibility—an episteme or way of knowing climate change—that allows for the reproduction of current practices and the continuation of certain types of political and economic structures. Mainstream media have some common traits such as large audiences, a commercial orientation, and a degree of proximity to official sources. Ideological cultures in the media can, of course, be significantly different, each with varying implications for how climate science and politics are represented (Carvalho 2007). But even the most progressive of dominant media build on certain—seemingly unquestionable—ideas that are at the foundation of climate injustice, namely the metadiscourses of free-market economy and elite policy-making, both underpinned by the metadiscourse of techno-scientific progress. The naturalization of such metadiscourses results from various interconnected journalistic choices, which, together with other problematic aspects in media practice, contribute to keeping societies locked into systems that produce inequality.

First, mainstream media have reproduced and endorsed the views of the most powerful actors in the political and economic spheres. They have typically ignored—and sometimes even discredited and de-authorized—social actors who challenge dominant value-systems. Several studies have shown that governments and intergovernmental organizations have strongly shaped media agendas with peaks in media attention in multiple countries having coincided with intergovernmental summits (e.g., Sampei and Aoyagi-Usui 2009). Governmental sources have been dominant in numerous contexts (e.g., Yun, Ku, Park, and Han 2012) while actors from the civil society have been routinely silenced.

Second, several media outlets, such as Fox News in the US, the *Daily Telegraph* in the UK, and *The Australian*, have chosen to offer a stage for denialist views on climate change, thereby precluding a healthy and informed discussion on injustice or on courses of action towards a more equitable and sustainable future. Painter (2011) and others have shown that this is mostly an Anglo-Saxon phenomenon with media in the US, Australia, and the UK being particularly prone to featuring denialists. Fossil fuel lobbies exacerbate the problem. The extent of the impact of the skeptic movement's public relations campaigns (led by wealthy think tanks and foundations, such as the American Enterprise Institute and the Heartland Institute) on American media cannot be overestimated, raising issues of representational equity and casting a shadow on the "public sphere" as a democratic discursive space. It must be noted, however, that in many other parts of the world the media tend to adhere to scientific findings on anthropogenic climate change (Painter 2011).

Third, the discourses that have tended to gain currency in mainstream media stay within the parameters of governance models that produce climate injustice. In-depth discourse analyses of media coverage have pointed to a frequent reproduction and legitimation of governmental discourses promoting techno-managerial approaches and to the marginalization of more transformative discourses (Carvalho 2011). The media do not function in a void. Social and political contexts and powerful institutions exert a strong influence on news reporting. Indeed, multiple discursive practices outside the media have contributed to the reduction of public debate on climate change. Already a decade and a half ago, Adger, Benjaminsen, Brown, and Svarstad (2001: 681) argued that a "global environmental management discourse representing a technocentric worldview by which blueprints based on external policy interventions can solve global environmental dilemmas" was dominant not only in climate change but also in the politics of deforestation, desertification, and biodiversity. How formal political bodies, given their privileged positions of power, frame the issue is key to its wider circulation. International organizations, such as the WTO, the IMF, and the World Bank (Methmann 2010), and governmental leaders, among others, have appropriated climate change in ways that serve the agenda of continuous economic growth with strong repercussions in media discourses (Carvalho 2005). The power relations that shape possible responses to climate change are absent from most media debates (although some exceptions should be noted, such as the fossil fuel divestment campaign led by *The Guardian*). Indeed, several scholars have argued that climate change is dominated by a "post-political" consensus (e.g., Swyngedouw 2010), with power issues, value assumptions, and choices being concealed. Expert-dominated consensus forecloses debate and often obscures injustice.

Fourth, there is a general tendency to suppress the ethical dimensions of climate change in mainstream media discourses. Research on the British press indicates that climate change is predominantly framed in terms that omit multiple ethical issues with presumptions of unlimited consumption and economic growth, for instance, remaining unchallenged (Carvalho 2005). Given the media's ultra-reliance on top political actors, this trend may be fed by politicians' silence on moral and ethical elements, as Gurney (2013) found in Australia.

A recent study of press constructions of climate justice pointed to striking differences between the US, India, and Germany (Schmidt and Schäfer 2015). Despite large variations between media outlets, a "freedom and resilience" pattern (privileging individual freedom and resourcefulness) was more commonly found in the US than in the two other countries. Taking into account the representation of the views of different social groups in the US press, Schmidt and Schäfer (2015: 545) argue:

In general, the compatibility of climate governance with market principles and the competitiveness of American industries, including effects on employment, are major concerns across all stakeholder groups. Only few references to "international solidarity" advance a different perspective, but they stem almost exclusively from foreigners.

In Germany, the study found mostly a "post-materialist debate on how to best protect values like environmental sensitivity and responsibility for distant others" whereas a postcolonial narrative (underscoring the right to economic and social development) was dominant in India (Schmidt and Schäfer 2015). Across the three countries, there was a pervasive nation-centric view of climate change and of options for action. While we should be wary of homogenizing analyses of mainstream media discourses and avoid reifying the mainstream-alternative dualism, it is fair to say that most dominant media have helped produce consent towards elite/technocratic decision-making and free-market capitalism, and steered well away from proposals based on climate justice principles such as fairness in the international distribution of commitments based on emissions per capita and historical responsibility.

Alternative constructions of climate change: accounting for plurality and climate justice

In contrast with the above, numerous "alternative media" have been developing discourses that provide different systems of intelligibility (or "ways of knowing" climate change) and promote other political subjectivities, thus offering a new hope for climate justice.

Atkinson and Dougherty (2006: 65) define alternative media as "any media that are produced by non-commercial sources and that attempt to transform existing social roles and practices by critiquing and challenging power structures." Although other scholars speak of different defining traits of alternative/independent/community media, overt engagement with social and political causes seems to be a common one. For example, *Democracy Now!* has remained committed to amplifying resistance to the construction of a major pipeline at Standing Rock Indian Reservation in North Dakota led by Indigenous people and a wider social/environmental movement (e.g., *Democracy Now!* 2016a, 2016b, 2016c). Hackett (2016: 14) has argued that the "most critical (in both senses of the term) functions" of alternative media are "counter-narrativity" and the "formation and mobilization of counter-publics." Here is how he envisages the former:

Counternarrativity entails filling in the gaps of dominant media accounts, finding the excluded voices and the dissonant facts that don't fit the official version, challenging repressive frames, providing new ways of making sense of contentious events and bringing attention to events and issues marginalised in the dominant media's topic agenda. (2016: 14)

Research on alternative media discourses on the environment and climate change is sparse but offers important insight on possibilities for counter-narratives. Comparing mainstream and alternative media coverage of metallic mining in El Salvador, Hopke (2012) concludes that the latter reframed the issue in terms of community rights and environmental justice. Similarly, Gunster (2012) analyzed two independent newspapers in Canada and found a significant emphasis on successful cases of action to mitigate climate change, which could act as exemplars for governments in enhancing civic pressure. He showed that political acts such as demonstrations, sit-ins and letter-writing campaigns were awarded much more space and salience than in the corporate media thereby opening up the politics of climate change to other actors. Examining several alternative Australian publications, Foxwell-Norton (2017) also found a critique of the politics of climate change, regular inclusion of voices that are otherwise marginalized, and an incitement to citizen action.

In an analysis of the claims of global movements for environmental justice, Schlosberg (2004) called attention to recognition as an integral part of the notion of justice and maintained that acknowledging and valuing diverse social groups, identities, and cultural practices is key to democratic environmental politics. As we have seen, strong roots of injustice grow in the dominant communication grounds/spaces through reproduction, naturalization, and legitimation of a narrow set of voices and discourses. To the extent that they offer discourses that are critical of hegemony and oppression, and pursue a politics of recognition (Fraser 1995) a more just future could be imagined in the horizon through alternative journalistic/communicative practices.

Such alternative journalism implicates news work and media organizations at multiple levels, including the all-important issue of funding (with several alternatives to corporate sources being experimented). While there is no single "recipe," journalism for climate justice thoroughly addresses inequities and suffering, as well as responsibility and agency, and sheds light on how societies can move forward towards sustainable futures. It puts a strong focus on those taking the brunt of climate change impacts, not only showing their condition but also discussing their rights, and constituting them into important political subjects in relation to climate change. It makes the sources of greenhouse gas emissions explicit: who, where, and how much are important questions to increase accountability and enable action. Enhancing citizens' sense of agency in climate politics can benefit from awarding visibility to and discussing possibilities of activism (Cross, Gunster, Piotrowski, and Daub 2015) and other forms of political engagement. Most importantly, climate justice-oriented media give more space to thoroughly transformative proposals to address climate change and its systemic causes, and contribute to repoliticizing climate change by promoting plural debate and open confrontation of ideas (Carvalho, van Wessel, and Maeseele 2017).

15 | SHIFTING THE BALANCE OF FORCES THROUGH SANCTIONS AGAINST TRUMP AND US CARBON CAPITAL

Patrick Bond

> Future climate scenarios are not only in the hands of state and corporate leaders; they depend upon the extent to which climate movement activists' current political philosophies, analyses, strategies, tactics, and alliances either weaken or strengthen the prevailing balance of forces. The most important barrier to reducing climate change remains Washington's philosophy, crudely expressed in 1992 when President George H.W. Bush told the Rio Earth Summit, "The American way of life is not up for negotiations." (Deen 2012)

In the same spirit, the Donald Trump administration removed the US from the 2015 *Paris Agreement* in June 2017 on the grounds that compliance will be too expensive for the world's largest economy (Trump 2017). In reality, starting with the Copenhagen Accord of 2009, Barack Obama's State Department ensured that United Nations climate negotiations were (unlike the Kyoto Protocol) voluntary and non-binding. The *Paris Agreement* avoided accountability mechanisms, and specifically prohibited "climate debt" liability lawsuits by climate victims for industrialized countries' prior pollution (Bond 2016), as even its chief negotiator Todd Stern (2017) brags. Yet in spite of Obama pledging only US$3 billion (in contrast to the several trillion dollars his administration spent on bailing out banks), Trump (2017) expressed misplaced concern about the United Nations Green Climate Fund "costing the United States a vast fortune," and that "massive liabilities" would result from damage done by US historic emissions.

Global-scale climate regulation by 2016 had become generally acceptable to the US population, even if many in support also voted for Trump. In November 2016, the Yale University Program on Climate Change Communication (Leiserowitz et al. 2016) poll of registered voters found that 78 percent supported taxing or regulating emissions, and 69 percent agreed this should happen in an international agreement. In 2009 even Trump publicly supported the Copenhagen Accord, although by 2012 he argued (on Twitter) that "The concept of global warming was created by and for the Chinese in order to make the US manufacturing non-competitive" (Trump 2012). His first 100-day plan stressed resurgent climate denialism as the default policy position: infrastructure construction focusing on fossil fuel pipelines, airports, roads, and bridges; cancellation of international obligations including withdrawal from Paris and default on payment obligations to the Green Climate Fund; retraction of shale gas restrictions; enabling the Dakota Access Pipeline and Keystone pipeline; purging of the Environmental Protection Agency (EPA); and a (futile) attempt to "save the coal industry." Further privatization of public land was also imminent, including Indigenous reservations, in search of more fossil fuels.

The retreat from Paris opens up a new opportunity for a revived strategy and tactic: *delegitimation of Trump, and sanctions against his regime and supportive US corporations more generally*. Formidable alliances could be ignited internationally with much more positive implications for climate futures than otherwise exist. Such "social self-defense" alliances (Brecher 2017) would ideally have been forged on the day of Trump's election in November 2016 (one network, United Resistance, appeared to do so in early 2017 but, aside from a www.unstoppabletogether.org website, did not actively unite the 50 progressive groups which signed up).

But so far, even after Trump walked out of Paris, these alliances remain only *potential* political approaches, because even the most sophisticated, militant US climate activists simply did not adopt *any* strategy aside from condemnation and defense of existing space (Funes 2017). There was no open discussion in the climate movements about how to change the balance of forces, aside from continuing to promote localized blockades against fossil fuel facilities, to defend (profoundly inadequate) state regulations and improve their enforcement, mostly via the courts, and to divest from the main fossil fuel companies and climate-destructive banks while encouraging reinvestment in clean energy. Each of these was a necessary strategy—but a much more decisive shift in the balance of forces will be necessary to secure a climate future that transcends just survival and moves society to the potentials Naomi Klein (2014a) discusses in *This Changes Everything*. Such post-capitalist visions include renewable community-owned energy, massive investments in public transport, the burgeoning of organic agriculture, compact eco-cities, a widely-shared green production ethos, humane consumption (so indispensable for the survival of the Global South), and "zero-waste" disposal so that oceans, rivers, and land may recover from the "Capitalocene" (Moore 2016).

Trump's survival requires a strategic rethink

The failure to take advantage of Trump's regime to ratchet up pressure reflects the US Left's general weakness. In spite of the political fragility, personal foibles, administrative chaos, leadership buffoonery, and shrinking legitimacy, Trump's first months in office failed to generate a sustained, unified response from the society's progressive forces. Most critiques by the local US and world Left came from specific incidents or from sectorally narrow interests. Protest marches on Washington regularly drew tens or even hundreds of thousands of women, tax justice advocates, scientists, and climate activists from January through April 2017, as well as impromptu immigrant protection rallies at airports. But these generally occurred without linkage or fusion, and without a convincing strategy for changing power relations. The most effective resistance to Trump came from either late-night comedians or competing elites.

However, there are important examples of powerful resistance, in part grounded in climate change advocacy. Activist groups – including Greenpeace, 350.org, BankTrack, and Sierra Club – which attacked the Dakota Access Pipeline owner Energy Transfer Partners and its creditors did "billions of dollars in damage" as a result of "campaigns of misinformation," according to the firm's lawsuit in August 2017 (Horn 2017). As a target of anti-corporate activism, according to 350.org's May Boeve and Brett Fleishman (2017: 1),

> Exxon is the most famous example because the company's own scientists actively studied the threat of climate change, and in response the company developed taller offshore drilling rigs in anticipation of rising sea levels. Yet while they were preparing for a warmer climate, they also funded campaigns claiming that the science was uncertain.

Exxon and other fossil fuel corporations were divestment victims of US$5 trillion in withdrawn stock market financing, thanks to thousands of activists in universities, pension funds, churches, and other institutions (Carrington 2016). City of London investment analysts Carbon Tracker had in 2012, recall Boeve and Fleishman, juxtaposed the amount of carbon the world could burn within "safe" limits of global warming and the amount of carbon embedded in the reserves of the publicly traded fossil fuel companies—the coal, oil, and gas planned for future production. It provided incontrovertible evidence that the companies intended to burn all this carbon, and against the backdrop of increased caps on doing so, thereby creating a high likelihood for a massive stock devaluation: a "carbon bubble." This attracted the attention of more mainstream investors, who began to rank the carbon bubble as a material risk. (2017: 2)

How far might this divestment movement reach into Trump's own wallet, and how far can his regime be delegitimated by a wider sanctions movement? Aside from repeated 2017 polls showing Trump with less than 35 percent support within the US, Pew Research (Wike, Stokes, Poushter, and Fetterolf 2017) pollsters reported in mid-2017 that much of the world is strongly anti-Trump. Most opposed are Mexico, Spain, Jordan, Sweden, Germany, Turkey, Chile, Argentina, Brazil, France, Colombia, and Lebanon, all recording their citizenries' support for Trump at less than 15 percent (only the Philippines, Vietnam, Nigeria, and Tanzania record more than 50 percent, although the two most populous countries, India and China, were not polled). Sanctions campaigning against rogue regimes is a time-tested approach that has often succeeded in the past. Especially in the event that Trump initiates yet another unjust US war, a "people's sanctions" strategy should put not only the President's and First Daughter's own product lines under pressure, but also tackle Trump-friendly big businesses such as ExxonMobil, Koch Industries, and Goldman Sachs.

Trump's vulnerabilities

Compared to any US leader in history, Trump's presidency offers a superb chance for a unifying campaign on climate, as well as other critical issues. By mid-2017 it was clear that the conservative-populist wave he appeared to be riding into office in late 2016—peaking in Britain with the June 2016 Brexit vote (or indeed in Hungary with Viktor Orbán's 2010 election)—had decisively ebbed. In August 2017 following the debacle of neo-Nazis openly marching in Charlottesville, Virginia, Trump's straightforward racist and fascist supporters were forced to retreat, both on the streets—in US cities such as Boston, San Francisco, and Seattle, as progressive activists vastly outnumbered the right—and in the Oval Office. The once-formidable alt-right influences of Steve Bannon,

Michael Flynn, Sebastian Gorka, and Rich Higgins were short-lived once the "Deep State" and mainstream media called them out (Rose 2017), leaving only Stephen Miller in place.

It soon became evident that, within the US, Trump failed to build a new right-wing coalition under paleo-conservative leadership (a term reflecting an "economic nationalist" orientation, in contrast to imperialist neo-conservatives). He also failed to take full control of the US state apparatus, and could not take advantage of the Republican hold over Congress. There, surprisingly high levels of disaffection were generated by two dozen "Republicans In Name Only" (Rinos, as pro-Trump alt-right critics called them), thus foiling health care cutbacks and other legislative initiatives. Trump's only genuine victories were appointing two reactionary members to the US Supreme Court and unravelling a generation of EPA environmental protection regulations, including rules on infrastructure construction that can withstand flooding.

The short-sightedness of this deregulation was exposed in the September 2017 hurricanes Harvey in Texas and Irma in Florida, whose intensity drew from the unprecedented warmth of Gulf waters. Also exposed was extreme differentiation in urban resilience along race and class lines (due to generations of ruling-class segregation strategies) as well as overall ecological vulnerability. Once 13 of 41 superfund toxic sites in Texas were flooded, toxic chemical fires erupted, 11 percent of US oil refining capacity was temporarily disabled, and the two hurricanes' US$200–300 billion in damages were calculated. The (Republican) mayor of inundated Miami, Tomás Regalado, begged, "This is the time that the president and the EPA and whoever makes decisions needs to talk about climate change. If this isn't climate change, I don't know what is. This is a truly, truly poster child for what is to come" (quoted in Friedman 2017).

Nearby, Trump's own Mar-a-Lago estate in Florida survived Irma, benefiting from a government flood insurance deal for rich coastal property owners. Three months earlier, in response to his withdrawal from the *Paris Agreement*, opposition members of Congress had proposed legislation to block such federal subsidies: the Prohibiting Aid for Recipients Ignoring Science (Paris) Act. Such climate-related delegitimation is vital for both internal and international resistance to Trump's regime.

Internationally, the would-be geopolitician Trump also failed to globalize his movement and identify logical allies for either building a climate-denialist front (he was alone in rejecting Paris, along with Syria) or for coming wars (e.g., against North Korea, Iran, and Venezuela—and perhaps later against China). His decision to deepen an ineffectual

16-year US quagmire in the Afghanistan War, his wild threats amidst nuclear brinkmanship in the Korean peninsula, and his weakness in Syria—in contrast to Vladimir Putin's strength of purpose—reflected a propensity to drop bombs indiscriminately on civilians, rather than identifying and pursuing substantive solutions. Trump's support for India against Pakistan, promotion of the feudalistic Saudi Arabian regime in intra-Gulf conflicts from Yemen to Qatar, and permission for ever more extreme Israeli Zionism, together confirmed his incompetence at managing the most volatile regions of the world—especially as the Middle East becomes increasingly tense and uninhabitable due to climate change. Likewise his natural allies fared poorly, as the Labour Party made surprising progress in the June 2017 British election and as fascist electoral threats anticipated in 2017 from Marine Le Pen in France, Geert Wilders in Holland, and the Alternative for Germany were contained.

But the most critical factor in his growing vulnerability would probably be the waning confidence Trump's capitalist class allies retained in his leadership. Immediately after Trump's Paris rejection, entrepreneur Elon Musk and Disney CEO Bob Iger quit his business advisory councils, as did several other leading managers of major corporations in August 2017 following his ambivalence about criticizing racists and fascists within his base, immediately after Charlottesville. To save face, he simply dissolved the two councils. Still, Trump's delegitimation was not complete, for important fractions of capital—especially in the real estate, construction, military, fossil fuel, and banking sectors—still anticipate much-improved profits if Trump's over-ambitious, carbon-intensive infrastructure program is launched and due to the massive tax cuts of 2018.

The next logical questions are whether Trump's weaknesses can be harnessed in aid of climate sanctions, and whether a route to eco-socialism can be identified from linking up a variety of progressive campaigns within climate justice. Specifically, in order to shift power to the extent necessary for such a transition, will a people's sanctions movement against the US elite also be necessary in coming months and years?

BDS-Trump advocacy

One immediate reaction to Trump's rise was a call for boycott and sanctions against his own firm and associates: Color of Change (2016) pulled Coca-Cola out of the 2016 Republican Convention sponsorship; Grab Your Wallet compelled Nieman Marcus, Belk, and Nordstroms to discontinue Ivanka Trump clothing sales; Sleeping Giants forced hundreds of advertisers which supported pro-Trump alt-right websites to withdraw their financing; and Boycott Trump has a long list of targets. Encouraged by the successes, a Boycott45 (2017) campaign expanded the sanctions strategies to Trump and Kushner tenant companies, on grounds their US$100 million in annual rental payments "enable and normalize Trump and Kushner's hateful and intolerant views and agenda, participate in Trump and Kushner's unprecedented lack of transparency to use the office of the President to enrich themselves, and strengthen Trump's political brand." High-profile Trump buildings are located not only across the US, but also in Istanbul, Seoul, Rio de Janeiro, Toronto, Vancouver, Panama, Uruguay, Manila, Mumbai, and Pune.

"Boycott Divestment Sanctions" (BDS) movements have recently been effective against Israeli apartheid and, during 1985–1994, can be credited with splitting white business from the South African apartheid regime, in conjunction with very strong local protest. BDS against the US could succeed *if US progressives are motivated to call for a world boycott of the US government plus key Trump-related corporations.* Implementing a BDS-Trump strategy will be an important challenge for climate activists the world over, argues Klein (2016b, 2017). She was soon joined by European Environmental Bureau leader Jeremy Wates (2017): "Trump is known to like walls. Maybe a wall of carbon tariffs around the US is a solution he will understand."

Indeed 25 major US corporations (including Apple, Facebook, Google, Morgan Stanley, Microsoft, Unilever, and Gap) warned Trump in an open letter that "withdrawing from the agreement ... could expose us to retaliatory measures" (Petroff 2017). Suddenly sanctions were discussed as a powerful, useful threat in diverse media sites like *Forbes* (Kotlikoff 2017), *Financial Times* (Wolf 2017), *DailyKos* (Lenferna 2017), *The Guardian* (Stiglitz 2017), and *The Independent* (Johnston 2017). The credibility of sanctions was enhanced by Nobel Economics Prize Laureate Joseph Stiglitz, who in a 2006 paper argued that, "unless the US goes along with the rest of the world, unless producers in America face the full cost of their emissions, Europe, Japan and all the countries of the world should impose trade sanctions against the US" (Stiglitz 2008). In May 2017, Stiglitz co-chaired a UN-mandated commission based at the World Bank that advocated widespread, urgent adoption of carbon taxes.

Even former French president Nicolas Sarkozy had in November 2016 raised the prospect of punishment against US products as a result of Trump's climate-destructive campaign promises: "I will demand that Europe put in place a carbon tax at its border, a tax of 1–3 percent, for all products coming from the US, if the US doesn't apply environmental rules that we are imposing on our companies" (Kentish 2016). A technical policy term for such sanctions emerged: "border adjustment taxes" or for short, border measures which avoid World Trade Organization anti-protectionist penalties (such taxes are not a "disguised trade restriction"). In a front-page story, the *New York Times* quoted a leading Mexican official at COP22 in Marrakesh just after Trump's win: "A carbon tariff against the United States is an option for us. We will apply any kind of policy necessary to defend the quality of life for our people, to protect our environment and to protect our industries," a point echoed by a Canadian official (Davenport 2016).

Ironically, when in 2009 Obama promoted carbon trading strategies within his ultimately unsuccessful pro-market legislative climate strategy, further incentives were discussed so that big corporations would agree to emissions caps. Establishment economists like the Peterson Institute's Gary Hufbauer and Jisun Kim (2009) observed that in such a context, US companies "paying to pollute" would need additional protection from outside competitors: "border measures seem all but certain for political reasons ... many US climate bills introduced in the Congress have included border measures ... [against] imports from countries that do not have comparable climate policies" (2009: 2–3).

Sanctions against a person (Trump), a power bloc (Trumpism), and a system (capitalism)

To ramp up the existing initiatives requires a major unifying effort by US progressive groups, and a realization that international solidarity will be a critical force in shifting the power balance. Making the process as democratic as possible is vital. In 2006, 170 Palestinian civil society groups initiated BDS, insisting on three unifying demands: the retraction of illegal Israeli settlements (a demand won in the Gaza Strip) and the end of the West Bank Occupation and Gaza siege; cessation of racially-discriminatory policies towards the million and a half Palestinians living within Israel; and a recognition of Palestinians' right to return to residences dating to the 1948 ethnic cleansing when the Israeli state was established. As BDS-Israel co-founder Omar Barghouti (2011: back cover) says:

> Boycott remains the most morally sound, non-violent form of struggle that can rid the oppressor of his oppression, thereby allowing true coexistence, equality, justice and sustainable peace to prevail. South Africa attests to the potency and potential of this type of civil resistance.

Ronnie Kasrils—a leader of the underground movement and from 2004–2008 the South African Minister of Intelligence—agrees:

> BDS made apartheid's beneficiaries feel the pinch in their pocket and their polecat status whether in the diplomatic arena, on the sporting fields, at academic or business conventions, in the world of theatre and the arts, in the area of commerce and trade and so on. Arms sanctions weakened the efficiency of the SA Defense Force; disinvestment by trade unions and churches affected the economy as did the termination of banking ties by the likes of Chase Manhattan and Barclays banks; boycott of products from fruit to wine saw a downturn in trade; the disruption of sports events was a huge psychological blow; dockworkers refusing to handle ship's cargoes disrupted trade links. (2015)

The strategy drove a wedge between the "English-speaking" Johannesburg capitalists and the "Afrikaner" Pretoria regime. As internal protest surged, it was the 1985 foreign debt crisis caused in part by BDS which broke the capital–state alliance and compelled South Africa's nine-year transition to democracy.

With Trump*ism* such a logical target, international solidarity to weaken that power requires a boycott of both high-profile state functionaries and key corporations in order to attack the legitimacy of profits made within a neo-fascist, climate-denialist US. As Public Citizen's Rob Weissman (2017) warns, the US faces "a government literally of the Exxons, by the Goldman Sachses and for the Kochs." In contrast, installing the ecosocialist governments required in the US and everywhere to generate a climate future that not only keeps the temperature within the scientifically necessary maximum and does so with *justice* at its very core will require a dramatic shift in the balance of forces. Such principles must be undergirded by further analysis of how to weaken the power structure, by the widening of delegitimation strategies beyond just Trump to major corporations, by the toughening of sanctions tactics, and by the forging of international alliances urgently required to repeat the South African BDS success.

16 | LINKING ENVIRONMENTAL JUSTICE AND CLIMATE JUSTICE THROUGH ACADEMIA AND THE PRISON INDUSTRIAL COMPLEX

David N. Pellow

Introduction

In a recent conversation Naomi Klein noted that one of the key failings of the mainstream environmental and climate movements has been their inattention to the centrality of how race, racism, gender, and patriarchy uphold the development and maintenance of capitalism, and the concomitant devastation of local and global ecosystems. As an example, she noted that, historically, some of the first subsidies to the fossil fuel industry were derived from the wealth extracted from colonized lands previously inhabited by Indigenous peoples and from the labor of enslaved African peoples in the Americas.[1] In other words, the core roles that white supremacy and patriarchy have played in producing the enormous threats to human and ecological health have been largely overlooked in much of the discourse and debate about climate change (see also Dauvergne 2017). The lesson I take from Klein's point is that it would behoove scholars and activists to do more intersectional analysis, theorizing, and organizing in ways that link issues, concerns, struggles, and solutions across seemingly disparate terrains. In this chapter, I link the field of Environmental Justice (EJ) Studies to the emergent field of Climate Justice (CJ) Studies by examining some links between and among universities, prisons, and EJ/CJ struggles. Because CJ and EJ are tightly interwoven discourses, practices, and visions of social change, I argue for connecting the institutions of academia and the prison system to CJ and EJ politics particularly in relation to divestment campaigns, because these institutions are instruments of social oppression and ecological harm. Specifically, I demonstrate that prisons and universities actively contribute to environmental racism and climate change through planning decisions and investments that support fossil fuel economies and place ecosystems and human health at great risk.

From environmental justice to climate justice and back

Environmental justice scholarship and politics have, for decades, emphasized the fact that many communities marginalized by economic, political, and social systems also face greater anthropogenic environmental threats (Bullard and Wright 2009; Lerner 2006). Climate justice scholarship and politics have documented the same, with a specific focus on climate-related hazards (Dunlap and Brulle 2015; Roberts and Parks 2006). The philosophies and politics of EJ and CJ, then, offer visions of a future where social justice, democracy, and ecological sustainability prevail. David Schlosberg and Lisette Collins (2014) demonstrate the myriad linkages and fusions between the movements and scholarship for EJ and CJ. For example, they note that concerns about climate change have long been present within EJ movements, with a particular emphasis on the goal of reducing fossil fuel extraction and carbon emissions, ensuring a "just transition" for workers and communities negatively impacted by a shift away from fossil fuel production. They also note that EJ movements demand accountability from fossil fuel industries for "fenceline communities" that are most heavily affected by petrochemical production (Lerner 2006; Shearer 2011). More recently, they point out, EJ movements have confronted the social forces that contribute to the vulnerabilities of communities hit by extreme weather events like Hurricane Katrina, which devastated much of the US Gulf region in 2005 (Bullard and Wright 2009, 2012). While EJ movements and EJ research have a longer formal history than CJ movements and research, the two have blended and integrated in many ways. As such "neither academics nor policymakers can comprehend the meaning of climate justice without understanding the long and pluralistic history of the social movements that have developed the concept over the past decades" and specifically that the "ideas, demands, and principles ... seen in the environmental justice movement ... had a direct influence on the conceptualization of climate justice" (Schlosberg and Collins 2014: 70). Both movements are multiscalar in their reach—from the local to transnational—and "demand attention—and challenges—to the existing relationships between human communities and the environments that sustain them" (Schlosberg and Collins 2014: 70). Furthermore, both movements have, at times, emphasized the need to challenge global capitalist systems and the concomitant risks that are imposed on marginalized communities, as well as emphasizing a desire to create mechanisms for ensuring procedural justice and compensation for those harms (Ciplet, Roberts, and Khan 2015; Tauli-Corpuz and Lynge 2008; United Nations WomenWatch 2009). In the next section I offer a sketch of two powerful social movements that might speak to key environmental, climate, and social justice concerns.

Two campaigns and an invitation to intersectional movement work

Fossil Free UC is a campaign led by students, faculty, staff, and alumni from the ten-campus University of California system that works to raise awareness of the hazards of the fossil fuel economy, to urge the university to divest from fossil fuels and allocate its investments with greater attention to climate risk and social responsibility (www.fossilfreeuc.org).

The Fossil Free UC campaign has had major successes since its inception in 2012. Former California Governor Jerry Brown was a UC regent and added his voice to the chorus of those calling for a change when, in 2014, he suggested that the UC study the possibility of divesting from coal. In September 2015, the UC system agreed to sell off its endowment and pension fund holdings in oil sands and coal companies—a decision worth US$200 million. UC officials cited "environmental concerns and rising financial risk" (Gordon 2015) in those sectors as driving reasons for that decision. They gave no credit to the ongoing UC-wide student movement that had specifically been pushing for this goal, but the role of activism was undeniable. However, divesting from tar/oil sands and coal still left traditional (non-tar sands) oil and natural gas investments untouched, so many Fossil Free UC student activists have taken up the call to "divest the rest!" Regarding the successful effort of the movement as well as the need for more work, UC Santa Cruz student activist Alden Phinney stated, "I think it's a really good move by the university. But it doesn't mean we are going to stop pushing for full divestment soon" (Gordon 2015). Toward that end, in the spring of 2017, students at several UC campuses took part in sit-ins to raise the stakes and visibility of their demand for total divestment from fossil fuels. Four UC chancellors endorsed their cause (Fossil Free UC 2017) and students are continuing work toward securing an agreement from the UC regents to achieve that goal.

In December 2015, just three months after Fossil Free UC achieved its first fossil fuel divestment success, the Afrikan Black Coalition (ABC)—also a University of California statewide student group—succeeded in pushing the UC system to divest $25 million from private prisons (Williams 2015). The Afrikan Black Coalition is a UC-wide campaign led by Black students concerned with drawing attention to and confronting the ways in which that university system's policies affect Black students and people throughout the African diaspora (afrikanblackcoalition.org).

This divestment from private prisons was prompted by concerns about the UC system's relationship to private prisons and to those who finance that industry, because of the disproportionate number of people of African descent who are incarcerated in such facilities. ABC Political Director, Yoel Haile, states:

> This victory is historic and momentous. Divesting $25 million is a good step towards shutting down private prisons by starving them of capital. This is a clear example of Black Power and what we can achieve when we work in unity. This victory belongs to the masses of our people languishing behind America's mass incarceration regime. (Williams 2015)

With this decision, the University of California became the second US educational institution to divest from private prisons, following a similar move by Columbia University in 2013, after a major campaign by Black students and their allies there. This means that the UC is the first public university to divest its shares and holdings from private prisons. The Afrikan Black Coalition pushed even further in announcing an additional demand that the UC must divest $425 million from Wells Fargo Bank. It called for this action because of the bank's "discriminatory lending practices in Black and brown neighborhoods" (Williams 2015); its role as a syndication agent and issuing lender for CoreCivic (formerly Corrections Corporation of America, a private prison company); its status as a trustee to the GEO Group (the largest private prison company in the US); and its membership in the Million Shares Club—a group of investors with at least one million shares in CoreCivic and the GEO Group; all of which meant "effectively financing the dehumanization of Black and migrant people" (Williams 2015). At the time of the UC announcement, ABC Field Organizer Kamilah Moore stated:

> In order for the UC's mission to be fulfilled, it is imperative to assess investments not only from a risk perspective, but from a socially responsible perspective as well. Our campaign is not over. We will continue to call for complete divestment, increased transparency, and reinvestment in education and businesses owned or controlled by the formerly incarcerated. (Williams 2015)

In both the Fossil Free UC and Afrikan Black Coalition campaigns we see a number of similarities and parallels. Both groups are statewide multi-university coalitions of students willing to take direct and public action to achieve progressive goals that exposed links between educational institutions and oppressive investment practices and the need for social change. They have achieved major successes through divestment campaigns that reorganized power dynamics between the UC regents and students. Having only achieved partial victories, both groups also declared their intentions to push further for more comprehensive changes.

However, what is disappointing about these campaigns is that it appears that there is no clear effort on the part of these two groups to work together and to link the divestment from fossil fuels to the divestment from prisons. In other words, the UC system is invested in both fossil fuel *and* prison industries, both of which perpetrate environmental, social, and climate injustices. Why not link these campaigns? I contend that activists and scholars can and should acknowledge the links between universities and prisons around the goals of racial, environmental, and climate justice because universities, prisons, banks, and the fossil fuel industry are instruments of social oppression and ecological harm. The other broader connection is that it is difficult if not impossible to separate racial justice from CJ and EJ, since they all are mutually reinforcing and because each is an example of the other. I explore these linkages in the next section.

The crucial link: prisons as sites of environmental and climate injustice

Environmental injustice and climate injustice are terms that scholars and activists use to describe the fact that environmental threats in general, and climate disruptions in particular, affect communities, nations, and regions of the globe differently and unevenly, with Global South communities, people of color communities, and Indigenous communities being hit the hardest (Ciplet, Roberts, and Khan 2015). These injustices in outcomes are compounded by the fact that these same communities have contributed the least to creating the problem of global ecological risks and climate change (Roberts and Parks 2006). It is these challenges that EJ and CJ scholarship and politics address. Fossil Free UC and the Afrikan Black Coalition have documented the fact that universities contribute to both environmental/climate injustices and the social injustices associated with the prison industrial complex (see also Sudbury 2009). But how, if at all, are these practices linked? The answer is simple: they are connected because prisons are sources and sites of environmental and climate injustice.

It is now clear that there are numerous ways that prisons and environmental threats intersect to produce harms to the bodies of prisoners and corrections officers and to nearby ecosystems (see Prison Ecology Project 2016). The following is just a sampling: (1) there are confirmed reports of water contaminated with arsenic, lead, and other pollutants at prisons in more than 20 US states, including the now infamous case of Flint, Michigan, where the Genesee County jail's inmates—including pregnant women—were forced to drink toxic water while prison guards drank filtered water; (2) the Rikers Island jail in New York City was built atop a landfill that for years has produced methane gas explosions and is plagued with health complaints from cancer-stricken corrections officers; (3) the Victorville Federal Correctional Complex in California was built on a former nuclear Weapons Storage Area (WSA) and is now a federally designated toxic military Superfund site (which means it is recognized by the federal government as contaminated land that poses a threat to human and/or environmental health); (4) the Northwest Detention Center in the Seattle-Tacoma area is a privately-operated prison designed to house more than 1,500 immigrant detainees and is built adjacent to a federally designated toxic Superfund site; (5) prisoners across many states recycle hazardous electronic waste for very low wages and enjoy few, if any, health and safety protections (Conrad 2011); and (6) Texas prisoners face some of the worst heat and humidity of any inmates in the nation, because of extreme weather combined with a lack of air conditioning and deficient medical care at facilities throughout the state. A recent investigation concluded that 14 prisoners died between 2007 and 2014 due to extreme heat and neglect by the Texas Department of Criminal Justice (Human Rights Clinic 2015).

The impacts of prisons outside their walls is also profound. Sewage overflows, chemical toxins, fossil fuel emissions, air pollution, and hazardous waste generated from inside many prisons affect ecosystems, nonhuman and human species and communities, as they also impact waterways, ambient air, and nearby land bases. Between 2006 and 2014, the Monroe Correctional facility (just north of Seattle) dumped half a million gallons of sewage into the Skykomish river, a popular recreational and fishing spot for people in this region (Anderson 2015). In another case, the proposed prison in Letcher County, Kentucky, to be built on a mountain top removal coal mining site near a low-income community, threatens at least 71 nonhuman species, many of them endangered (including the Indiana bat and the grey bat) (Tsolkas 2015).

The fact that these nonhuman species and ecosystems are more vulnerable via the prison system speaks to the ways that human and nonhuman communities are linked through what David Nibert (2002) calls "entanglements of oppression." Communities are entangled within and across species in ways that reveal how the scholarship on environmental inequality and intersectionality can benefit from extending beyond the restrictive boundaries of the human to encompass and observe the modes through which humans and nonhumans are connected through discourses, policies, and practices of oppression and privilege making. And while universities are sites where *investment* decisions contribute to climate change and the prison industrial complex, academia is also complicit in that many scholars develop *research and promote ideas* that support strengthening the criminal legal system and promote the denial of environmental racism and climate change (Been 1994; Dunlap and McCright 2013; Sudbury 2009). For example, Roy W. Spencer is a research scientist at the University of Alabama, Huntsville who has received funding from the coal company Peabody Energy to publicly testify that humandriven climate change is nonexistent. In another example, Dr. Edward Schein is infamous among activists in the anti-prison community for having developed theories of brainwashing and behavior modification that amount to deprivation and torture. His 1961 presentation at a meeting convened by the Bureau of Prisons led to the widespread adoption of his techniques in prisons throughout the US (James 2003: 191–3). Thus, because prisons and academia are spaces where environmental and climate injustices are evident, amplified, and ongoing, I suggest that the Fossil Free UC and Afrikan Black Coalition might see fit to connect their campaigns and visions for social change.

Discussion and conclusion

Articulating productive linkages between EJ and CJ scholarship and politics, focused on the unequal burdens of anthropogenic socioecological violence, is necessary because they are both tightly interwoven discourses, practices, and visions of social change. These linkages can be realized in a material fashion by connecting struggles for fossil fuel divestment with divestment from the prison industrial complex.

Urban political ecologist and critical geographer Nik Heynen builds on W.E.B. DuBois's concept of "abolition democracy" to formulate the idea of "abolition ecology," which I find useful here. Abolition ecology is "an approach to studying urban natures more informed by antiracist, postcolonial, and Indigenous theory. The goal of abolition ecology is to elucidate and extrapolate the interconnected white supremacist and racialized processes that lead to uneven development within urban environments" (Heynen 2016: 839). Heynen's concept is rooted in a critique of the logics of racial capitalism, colonialism, and the plantation economy and their ongoing impacts on society. While I find Heynen's concept generative, in the context of the prison/university-industrial complex, I would extend it to embrace a focus on: (1) rural environments, since many prisons and modern day plantations are in rural spaces (see Braz and Gilmore 2006; Gottlieb and Joshi 2013; Tsolkas 2015); (2) a call for the abolition of *all* forms of hierarchy and domination, not just or primarily those that involve or derive from the enslavement and colonization of human beings; and (3) supporting campaigns of nonviolent direct action to achieve the above. This approach is informed by what I have called "total liberation" (Pellow 2014) and critical EJ (Pellow 2016), which constitute a call for the abolition of the nature/culture hierarchy and human–nonhuman dominionism, as well as the abolition of state and corporate authoritarian formations that support those forms of violence. If, as Angela Davis (2005) has argued, the abolition of prisons is the unfinished work of the abolition of human slavery, then abolition ecology, critical EJ studies, and total liberation expand and deepen that work to include an effort to radically undo and transform all known forms of unfreedom. Such an approach would argue for the abolition of inequality within human societies and between humans and the more-than-human world and across all spatial formations (including urban, rural, aquatic, aerial), a move that would seem to be a logical extension of the abolition democracy and abolition ecology approaches.

One of the most important accomplishments of the EJ movement was the drafting of the Principles of Environmental Justice (see www.ejnet. org/ej/principles.html)—a vision of the future that is radically different from the present and involves deep transformations within human societies and in our relationships with our multispecies world. The evidence suggests quite clearly that prison systems are entirely incompatible with a world in which social, climate, and environmental justice prevails. If the abolition of prisons requires the building of relationships across society and *away* from prisons that disrupt the sources of power that fuel the carceral system, there is a need to encompass a multispecies EJ approach. Such an approach would confront *multiple* forms of dominance within and beyond the boundaries of the human species to support relationships, knowledge, traditions, and practices that are sustainable and *just* in order to render the prison industrial complex in particular and environmental and climate injustice more generally unacceptable, irrelevant, and past tense.

There is now ample evidence that both universities and prisons are sites where the struggle against environmental and climate injustices needs to be waged. I have taken the liberty of offering unsolicited advice to the Fossil Free UC campaign and the Afrikan Black Coalition to consider building enduring connections between their efforts to create change inside and outside of academia and the prison industrial complex. This is a challenge to both movements to notice and act on the historical and contemporary connections among the fossil fuel and prison industries and the socially and environmentally oppressive practices that constitute the core of their foundations and the core logic of global capitalism itself. In other words, drawing these connections is not an exercise in abstract theory; it is a cultural and material necessity for our survival.

Note

1. Conversation with faculty and students at University of California Santa Barbara, May 17, 2017.

17 | DEMOCRACY AND CLIMATE JUSTICE: THE UNFOLDING OF TRAGEDY

Walter F. Baber and Robert V. Bartlett

The climate change narrative has entered contemporary political discourse in a Janus-faced form. On the one hand, the climate warming narrative is a discourse of judgment, pathology, and catastrophe. It is a hostile "other" in collective consciousness, condemning us to suffer for our excesses, threatening us with environmental conditions that Western culture has long associated with ill-health—heat, humidity, miasma, languor, decay—and holding over our heads the prospect of a radical dispossession of everything that we value (Hulme 2008).

On the other hand, climate change in the hands of the global elite has become an "empty signifier" (Methmann 2010). An analysis of discourses of the WTO, IMF, World Bank, and OECD suggests that the global governmentality of climate protection is built on four discursive pillars—globalism, scientism, efficiency, and an ethics of growth—that make climate protection function as a powerful but meaningless rhetorical tool. It presumes that it is possible to integrate climate protection into the global hegemonic order without changing the basic social structures of the world economy or disrupting business as usual. International organizations of all sorts can claim to be in favor of climate protection without accepting any real obligation while simultaneously arguing for significant limitations on the behavior of those who inhabit particular places and occupy a lower rung on the socio-cultural ladder (Methmann 2010). Again, the climate narrative is received as a hostile other, seeking to impose its alien order of the values of globalism, scientism, efficiency, and growth while self-servingly exempting itself from the implications of those values—prompting both climate denialism in general and accusations of hypocrisy against specific elite sources of climate warming discourse.

So, here we see the playing out of tragedy, in the fullest literary or theatrical sense, evident in the interweaving of the avoidable and the inevitable in the collective hubris and self-delusion that constitutes modern climate politics. The Dionysian sense of climate warming as a threat of deprivation, death, and damnation is married to the Apollonian smugness, superiority, and separateness of global environmentalism. Is it any wonder that democratic processes that rely only on the majoritarian aggregation of shallow, tentative opinions have trouble pursuing climate protection without suffering cultural backlash? Not every voter is equipped with the sense of balance and commitment to reasoned judgment that is required to withstand an assault on their personal values by what often seem to be fire and damnation preachers in crisp, white lab coats. The view that ultra-nationalism is a reaction primarily to economic deprivation and political disenfranchisement is not well-supported empirically, but it is a convenient retreat for those who view climate politics as merely a matter of striking the right bargain in some fundamentally rational negotiation.

To the contrary, climate politics is a central element of a broader process that is roiling the developed world. Since about 1970, relatively affluent and generally democratic societies have seen "a growing emphasis on post-materialist and self-expression values among the younger birth cohorts and the better educated strata of society" (Inglehart and Norris 2016: 29). This has led not only to a greater concern for environmental protection, but also an increased acceptance of gender and racial equality and equal rights for LGBT peoples. This broad cultural shift has fostered greater approval of social tolerance of diverse lifestyles, religions, and cultures, multiculturalism, international cooperation, democratic governance, and protection of fundamental freedoms and human rights. Social movements reflecting these values have brought policies such as climate warming to the center of the political agenda, drawing attention away from the classic economic redistribution issues. But this spread of progressive values has also stimulated a "cultural backlash" among people who feel threatened by this development. Less educated and older citizens, especially white men, who were once the privileged majority in Western societies, resent being told that traditional values are "politically incorrect" and have come to feel that they are being marginalized within their own countries. As this cultural backlash has continued, a "tipping point" appears to have been reached in support for populist parties and leaders who "defend traditional cultural values and emphasize nationalistic and xenophobic appeals, rejecting outsiders, and upholding old-fashioned gender roles" (Inglehart and Norris 2016: 29–30).

Scholars who study comparative environmental politics find strong empirical support for the conclusion that nondemocratic societies can never develop the kind of cognitive, cultural, epistemic, institutional, and economic capacity that ultimately must be required for sustainable climate governance (Christoff and Eckersley 2011; Dryzek, Norgaard, and Schlosberg 2013). Purely aggregative majoritarian political systems have performed better, and are likely to continue to do so, although none yet has cultivated capacity remotely adequate to the challenge (Stevenson and Dryzek 2014). Such "vote-centric" (Chambers 2003: 308) democracy is fundamentally limited, combining as it does "the 18th century procedure of voting with the 19th century idea of universal suffrage, the 20th century invention of mass media, and the 21st century culture of social media" (Van Reybrouck 2016).

Achievement of effective, ecologically rational climate governance is unlikely ever to be facilitated by aggregative, majoritarian democratic practices alone, even if serious effort were undertaken to extend and guarantee the franchise, broaden and deepen authenticity, widen the scope of control, and mitigate the various inequalities of contestation (Dryzek 1996; Stevenson and Dryzek 2014). Even a generally well-educated population will have persistent trouble avoiding the misinformation, manipulation, and exploitation that are certain to always be part of the mass symbolic politics of majoritarian aggregative democracy. Both the advanced state of information and communication technology and the stunted quality of journalism magnify the problems and accelerate the attendant alienation that is inexorably hollowing out the legitimacy of any kind of vote-centric democracy.

But there is considerable environmental promise in democratizing governance in ways that center talk, listening, and reflexivity—a promise that far outweighs the perils attendant to that sort of deliberative democratic practice. Of course, such deliberative democracy—governance by discussion—even if fully adequate from a political perspective, can still produce ecologically irrational results (Bartlett 1986; Dryzek 1987). Yet the most ecologically sophisticated policies imaginable will prove unsustainable if they fail the test of democratic legitimacy (Baber and Bartlett 2005, 2009). Thus deliberative democracy is a necessary (though never sufficient) element of environmental sustainability and, more importantly, it is an essential element in any solution to the problem of cultural backlash against climate protection.

Questions regarding inclusion and representation abound in deliberative democracy and are central to the problem of cultural disenfranchisement.

Inclusion has not just intrinsic value but instrumental value, increasing the likelihood of making better or correct choices (Stevenson 2016). The discursive character of deliberative practice suggests quite clearly that what is important to include are the narratives of all, rather than the votes of all. In other words, a deliberation that in any way fails to include any meaningful narrative is democratically deficient. In pursuit of the goal of inclusiveness, it may make sense sometimes to modify some of deliberative democracy's operating rules of thumb. For instance, a diverse range of participants is thought to be vital to produce deliberative results of value. Where politically disadvantaged populations are concerned, however, the effective development of their narratives may require (at least preliminarily) enclave deliberation that allows participants to develop, assess, and refine their own narratives in a relatively homogenous environment before exposing them to the rigors of the marketplace of ideas (Karpowitz, Raphael, and Hammond 2009). But where a sense of cultural disenfranchisement is a problem, an essential objective of deliberative design must be an exposure to the "other" designed in such a way as to promote the awareness of shared values and perspectives.

Likewise, with regard to the importance of integrating "traditional knowledge" and perspectives into environmental decision-making, the last word has not been said—nor is ever likely to be. That is because to be deliberatively effective, knowledge (lay or expert) must be not merely local, but culturally embedded. Recent field research suggests that the development of democratic deliberation depends more on whether participants situate and link their knowledge than whether the knowledge is lay or expert in origin. This suggests that embedded (or grounded) knowledge, situating one's experiences in a way that enables participants to actively link with other sources of knowledge, is a more useful concept for anyone who wishes to better understand which ways of knowing enable environmental deliberation in participatory processes (Ashwood, Harden, Bell, and Bland 2014). This kind of embeddedness is imperfectly understood—in part because what it requires is, and is likely to always be, context specific to a very considerable degree.

As with empowerment and embeddedness, the demands that equity places on deliberative environmental democracy also need to be explored more thoroughly. For example, the pursuit of environmental justice introduces both problematic participants and problematic relationships to deliberative environmental democracy (Baber and Bartlett 2005). The use by environmental justice advocates of their own unique storylines can be an important mechanism for shaping policy meanings and for improving deliberative quality, although these effects are tempered by discursive and material forms of power and the competition among alternative storylines (Baber and Bartlett 2009; Dodge 2014). A new challenge in this regard will be to discover deliberative mechanisms for extending deliberative environmental democracy techniques to the analysis of international equity problems for which they were not originally intended (Baber and Bartlett 2009, 2015) such as the stubborn gridlock surrounding global climate politics (West 2012). Obviously, the differentially distributed consequences of climate change are likely to be a primary source of these alternative story lines, and exposure to them in an adequately deliberative context carries significant potential to counter the forces of cultural backlash. Likewise, another such frontier is the development of deliberative environmental democracy principles and practices that will allow both scholars and citizens to explore problems of intergenerational justice in ways that are both more practical and more theoretically defensible (Cotton 2013). The implications of this emerging narrative for climate politics are self-evident.

Many of our discussions of the peril posed to deliberative environmental democracy by the risk of elite cooptation suggest that additional thought needs to be given to what it means to call environmental policy effective. If environmental decision-making meets the criteria we have identified so far (if it is empowering, embedded, and equitable), then its effectiveness could only be degraded if it were coopted by self-serving elites. This will strike some environmentalists as deplorably anthropocentric. In some ways, it is. But that very accusation is growing increasingly untenable. If there is a cutting edge to global environmental governance understanding, it involves the concept of the "Anthropocene" (Biermann et al. 2012). This concept suggests that no part of the natural world today is untouched by humans and, therefore, no solution to environmental problems can avoid placing humans near its center. But even this may understate the case. Humanity today is so omnipresent that the very nature of nature has been altered (Wapner 2014). The distinction between the human and nonhuman components of nature, one that environmentalists have long used to justify both wise use and preservationist policies, is no longer tenable. Today, there is only a distinction between the human and the more-than-human environments. In the Anthropocene, environmental protection demands expansive, reflexive listening to the environment, "attuning ourselves to the hybrid character of ecosystems and helping to shape them in ways in which the human voice is *deliberately* one among others fashioning socio-ecological arrangements" (Wapner 2014: 46, emphasis added). Deliberative democracy's historical commitment to consensus may have been too narrow rather than over-broad (Baber and Bartlett 2015). The same can easily be said of democratic practice generally, and the consequences of that narrow expectation regarding democratic consensus may have finally taken human form on the far right.

A paradox about consensus is alleged to lie at the heart of deliberative democracy and its implications for the future progress of governing human actions affecting climate. It is entirely plausible, at least theoretically, that consensus-oriented political practices could eventually fall victim to the same sort of political decay that plagues their aggregative relatives (Fukuyama 2014), ultimately leaving the processes of climate change diminished, perhaps, but to a wholly inadequate extent. Although the symptoms of political decay in these two cases might appear similar, the underlying causes would be very different. Samuel Huntington's (1965) original conception of political decay was based on the insight that political and socioeconomic modernization leads to the mobilization of new social groups over time whose new demands cannot be accommodated by existing political institutions. In the case of deliberative environmental democracy, however, other factors might work to some degree at cross purposes. The danger might be, for example, that environmental policies that are *effective* (in part because they are *experimental*) would eventually have their effectiveness undermined precisely because they were *empowering, embedded,* and *equitable.* In the present context, the challenge of climate change would remain unmet not because politics was insufficiently democratic but, rather, because it had become too democratic. Just as unending analysis can be the cause of political paralysis and ineffectiveness, so can unbounded deliberation result in decision by non-decision. About this danger, at least two observations are possible. First, Huntington's analysis suggests that our concerns about political decay should not lead us to abandon deliberative environmental democracy because none of its competitors are capable of producing institutional arrangements that are more lasting. Second, to the extent that deliberative environmental democracy does produce decisions that are genuinely consensual, the problem of political decay has been significantly simplified. If the source of political decay is not to be found in our political stars, but in ourselves, then the remedy for decay is within us as well. What is required is, merely, a more *experimental* understanding of consensus itself. Any more-than-superficial consensus, like all real learning (Dewey 1916), must be grounded in experience, in broadly democratic engagement. Consensus can be organized by an appreciative inquiry into the problem-solving capacities revealed by people in concrete circumstance—not an uncritical acceptance of whatever seems to be the popular solution, but a search for the principles that are potentially universalizable in the Kantian sense—as the guidance that we would want others to apply to us in our own experiences.

So, climate governance can be achieved only by consensus, and yet consensual climate governance will only be possible if there is also a concurrent emerging consensus that the ways humans begin to govern themselves as they affect the planetary climate is resulting in some significant and notable "removal of manifest injustice in the world" (Sen 2009: 7). Successful climate governance must be seen to be reducing injustice, not increasing it or only affecting it indifferently. Just as deliberative democracy is a necessary (though never sufficient) element of environmental, and thus climate, sustainability, the same processes of public reasoning are necessary for both identifying redressable injustice and pursuing its redress.

> The crucial role of public reasoning in the practice of democracy makes the entire subject of democracy relate closely with ... justice. If the demands of justice can be assessed only with the help of public reasoning, and if public reasoning is constitutively related to the idea of democracy, then there is an intimate connection between justice and democracy. (Sen 2009: 329)

Successful and effective climate governance necessarily requires a considerable degree of deliberatively achieved consensus among all peoples. Achieving that consensus necessarily will depend on the simultaneous emergence of a consensus understanding that climate governance is moving the world toward re-imagining and seriously addressing the demands of climate justice.

The climate choices that global humanity faces are hardly challenging, either technically or morally. But the modern human capacity for hubris and self-delusion is as great, if not greater, than ever was dramatized in classical Greek theatre. Hope remains, but this late in the last act, there is little justification for optimism. Disaster can be avoided, yet humans insistently remain on paths that will make inevitable the unfolding of the tragedy of democracy and climate justice.

PART FOUR

THE QUEST FOR CLIMATE JUSTICE ACROSS THE WORLD

18 | THE MAJURO DECLARATION FOR CLIMATE LEADERSHIP: A PACIFIC APPROACH TO A GLOBAL PROBLEM

Lagipoiva Cherelle Jackson

Introduction

The *Majuro Declaration for Climate Leadership* (Pacific Islands Forum 2013a), an initiative of Marshall Islands, is a unique approach by Pacific Island nations to take a collective stance on climate change outside of the prescribed international negotiations under the *Kyoto Protocol.*

Pacific Island countries have a population of about 3.4 million people, spread across hundreds of islands, and scattered over an area equivalent to 15 percent of the globe's surface. This unique and diverse region is one of the most vulnerable parts of the world to the effects of climate change and natural disasters. Based on a World Bank (2017) report, of the 20 countries in the world with the highest average annual disaster losses scaled by gross domestic product, eight are Pacific Island countries: Vanuatu, Niue, Tonga, the Federated States of Micronesia, the Solomon Islands, Fiji, Marshall Islands, and the Cook Islands.

Climate change is already disproportionally affecting the islands of the Pacific. Although islanders have done little to contribute to the cause—less than 0.03 percent of current global greenhouse gas emissions—they are facing the devastating impacts of climate-induced sea level rise, sea temperature increases, ocean acidification, altered rainfall patterns, and overall temperature on communities, infrastructure, water supply, coastal and forest ecosystems, fisheries, agriculture, and human health (Secretariat of the Pacific Environment Programme 2017). The Pacific, particularly several of the low-lying coral islands such as Kiribati, Tuvalu, Marshall Islands, and Tokelau, is in fact without a doubt most at-risk from natural disasters (Solomon et al. 2007). Climate change is already affecting the islands with dramatic revenue loss across sectors such as agriculture, water resources, forestry, tourism, and other industry-related sectors (IFAD and The Global Mechanism 2009). Catastrophes including large-scale flooding, erosion, and intrusion of sea water would result in economic and social costs beyond the capacity of most Pacific Island countries and threaten the very existence of small atoll countries. A rise of average sea level by just one meter, when superimposed on storm surges, could easily submerge low-lying islands.

Pacific Islands Forum

The political platform for Pacific Island governments to respond to disasters that relate to climate change among other catastrophes is the Pacific Islands Forum hosted through the Pacific Islands Forum Secretariat (PIFS). This is a political grouping of 16 independent and self-governing states, founded in 1971 as the South Pacific Forum. Members include Australia, Cook Islands, Federated States of Micronesia, Fiji, Kiribati, Nauru, New Zealand, Niue, Palau, Papua New Guinea, Republic of Marshall Islands, Samoa, Solomon Islands, Tonga, Tuvalu, and Vanuatu. New Caledonia and French Polynesia, previously Forum Observers, were granted Associate Membership in 2006. Current Forum Observers include Tokelau (2005), Wallis and Futuna (2006), the Commonwealth (2006), the United Nations (2006), the Asian Development Bank (2006), Western and Central Pacific Fisheries Commission (2007), the World Bank (2010), the ACP Group (2011), American Samoa (2011), Guam (2011), and the Commonwealth of the Northern Marianas (2011), with Timor Leste as Special Observer (2002). The annual Forum meetings are chaired by the head of government of the host country (currently Cook Islands), who remains as Forum Chair until the next meeting. Since 1989, the Forum has held Post-Forum Dialogues with key dialogue partners at ministerial level. There are currently 14 partners—Canada, People's Republic of China, European Union, France, India, Indonesia, Italy, Japan, Republic of Korea, Malaysia, Philippines, Thailand, United Kingdom, and the United States.

All members of the PIFS are among the 149 signatories to the *Kyoto Protocol*, an international agreement linked to the United Nations Framework Convention on Climate Change (UNFCCC), which commits its parties by setting internationally binding emission reduction targets. The Conference of the Parties to the *Kyoto Protocol* (COP) takes place annually and is the supreme decision-making body of the Convention. Pacific Island governments negotiate under several groupings which include the Alliance of Small Island States (AOSIS), Group of Least Developed Countries (LDCs), and G-77 and China. Due to the complexity of the negotiations, entrenched alliances within these groupings, and limited negotiation capacity of Pacific Island governments, regional representatives struggle to assert their voices during the negotiations. Despite the fact that two Pacific Island countries now hold leadership roles in the major groupings with Nauru being the head of AOSIS and Fiji the Chair of G-77, efforts to be recognized at the negotiations are still challenging for Pacific governments.

Majuro Declaration for Climate Leadership

On September 5, 2013, the Pacific Islands Forum, at its meeting in Marshall Islands, adopted the *Majuro Declaration for Climate Leadership* (Pacific Islands Forum 2013a). The Declaration highlights the Pacific's political commitment to be a region of climate leaders and its effort to spark a "new wave of climate leadership" (Pacific Islands Forum 2013b) that accelerates the reduction and phasing down of greenhouse gas emissions. It is also the first text of its kind to encourage commitments from both governments and non-state actors, including cities, companies, and other organizations, and is intended to complement and build momentum under the UNFCCC at a crucial time.

According to the Declaration the situation could not be more urgent. Waiting for a new global agreement will not be enough. Accelerating climate action before 2020 is critical. This is the urgency of now (Pacific Islands Forum 2013b).

The 12-page document says governments in the region are committed to demonstrating "climate leadership" and calls on countries to list "specific" pledges to reduce pollution.

To summarize, the Declaration:

Recognizes the gross insufficiency of current efforts to tackle climate change, and the responsibility of all to act urgently to reduce and phase-down greenhouse gas pollution; Confirms the Pacific Islands Forum's climate leadership in the form of their ambitious commitments to reduce emissions and the significant benefit in transitioning to renewable, clean and sustainable energy, and their desire to do more with the cooperation and support of international partners; and Calls on others—in particular Post-Forum Dialogue Partners, but also other governments, cities, the private sector, and civil society—to commit to be Climate Leaders by listing specific commitments that contribute more than previous efforts to the urgent reduction and phase-down of greenhouse gas pollution.
(Pacific Islands Forum 2013b)

The *Majuro Declaration* is also a dynamic document, which strongly encourages governments to continue to scale-up their action on climate change. According to one commentator the spirit of the agreement is to reframe the "I won't move till you move first" stance many world governments continue to hold, to an "I'm moving and I invite you to move with me" (Rigg 2013) one. According to the same commentator, if more countries follow this lead, it would totally change the dynamic within the UN climate negotiations (Rigg 2013). Marshall Islands President Christopher Loeak said he hoped the Declaration could be a "game changer" in driving talks on a global emissions reduction deal forward: "We need the rest of the world to follow the Pacific's lead" (King 2013). Similarly, the Pacific Islands Forum Secretary General Tuiloma Neroni Slade believes the Declaration to be "a declaration of responsibility" in which Pacific countries can "demonstrate their leadership" in mitigating climate change (RNZ 2013).

Implications of the Declaration

The Declaration has several implications but I specifically discuss two in this chapter, one a positive implication focused on the proactive and unique nature of the initiative, and the other a negative implication based on the risks taken by the Pacific by signing on to this declaration.

The Declaration is the first of its type, bringing together governments to work towards tangible action in a partnership that is not part of an international process or a United Nations convention. It demonstrates solidarity between small island nations of the Pacific on an issue that threatens the survival of some islands. It shows to the international community that Pacific Island governments are willing to act outside of internationally mandated political responses in order to find a solution to the problem, and, perhaps critically, that it is willing to work with partners outside of the political realm, which include the private sector, civil societies, and others. "The responsibility of all to act falls to every government, every company, every organization and every person with the capacity to do so, both individually and collectively" (Pacific Islands Forum 2013b). The text underlines the intense frustration among leaders of small island states at the sluggish progress at the UN in cutting global greenhouse gas emissions (King 2013).

On the other hand the Declaration may have a negative impact on the Pacific engagement at the international negotiations, and ultimately may backfire in the region's attempts to be considered as a legitimate voice at the UNFCCC negotiations because of perceptions of disrespecting the international process and reinventing the wheel. Although honorable in its intention, the Declaration could inadvertently signal concrete steps by regions to disregard international processes and focus on their own regional approaches and potentially make way for some governments to make excuses for not complying with non-binding international treaties and regulations not limited to climate.

Conclusion

Published media articles and high-level statements by leaders suggest that the *Majuro Declaration* has proven to be a successful political move for the Pacific which has led to extensive publicity of Small Island Developing States engagement in climate change issues.

It has now been four years since the Declaration was announced by the Pacific Islands Forum, and there is little evidence that it has influenced international climate negotiations or led to united or high-level action at the UNFCCC negotiations by Pacific Island countries.

The *Majuro Declaration* although it falls short of global significance, has offered a platform for some Pacific Island countries to strengthen their Intended Nationally Determined Contributions (INDCs), a process under the UNFCCC that requires countries to declare their national actions. Kiribati and Tuvalu have referenced the Declaration in their INDCs as a supporting factor in their national actions.

Although the Declaration was a positive move to declare political will by Pacific Island leaders towards action on climate change, it remains an expression of goodwill towards a critical issue globally but not a catalyst for change, as many may have hoped. Due to the non-legally binding nature of the Declaration, it is yet another commitment that follows the long list of politically motivated high-level agreements that do not have a substantive impact on climate change solutions, or international negotiations. It does, however, provide a platform for Pacific Islands to express their own commitments at the national level, and offers solidarity amid the pursuit of climate solutions for the islands of the Pacific. What the *Majuro Declaration* has proven is that at the very highest level of leadership in Pacific Island states, there is a willingness to address the issue to ensure the safety of the people, cultures, and land of the Pacific.

19 | COMMUNITY APPROACHES TO CLIMATE JUSTICE: CASES FROM PAPUA NEW GUINEA

Sangion Appiee Tiu

Our world today is faced with the challenges of deteriorating environmental conditions associated with the changing climate. Changes associated with climate change have varied effects across the globe, with those most affected being people and communities who are vulnerable, with little or no means to address the problem. Frequently encountered changes include prolonged drought, continuous rains, flooding, landslides, sea level rise, and coastal erosion.

For Small Island Developing States (SIDS) the challenges are many and come in all forms. For example, in Fiji, the frequency of cyclones and strong winds that lead to floods, landslides, and power cuts is increasing (Fox, E. 2016). In fact, News Corp reported that in April 2016, Fiji was struck by Tropical Cyclone Zena just a few weeks after being devastated by Tropical Cyclone Winston (Payne, G. 2016), demonstrating the deepening impact of the changing climate on low-lying island states. Other very small island nations such as Tuvalu and Kiribati with many low-lying atolls and islands are also threatened by the rising sea levels, with water flooding schools, cemeteries, roads, and villages at king tides. These examples demonstrate that solutions and responses would require the cooperation of all stakeholders to address climate-related disasters and their impacts. Every citizen and resident of these island nations has the responsibility to collaborate in order to strategize and take actions. The actions could be either to empower community capacity to develop resilience for climate-related disasters or, to take adaptive measures to minimize the effects on people and environment.

In Papua New Guinea (PNG) the increasing number of climate-related disasters in the interior areas of the country such as flooding of villages and food gardens due to heavy rains and continuous landslides greatly affect the delivery of services (Davies 2016; International Organization for Migration 2016). There, civil society organizations have recognized the plight of the people and have initiated climate actions in various ways to help communities either minimize the impact of climate change or adapt to the changes that have occurred. This chapter shares three examples of such organizations.

Case 1: Community initiative for immediate action— Tulele Peisa Inc.

In the Carteret Islands, a *Carterets Integrated Relocation Program* was initiated by the Carteret Islanders when they recognized the threats caused by the rising sea level. Throughout the islands, many coconut palms and shade trees on the beach continued to fall as the water level rose. Many of their islands lost most of the land, such as Huene Island, which had over 50–60 percent of its land washed away by the rising sea level since the 1980s (Rakova 2012). In addition, gardens were inundated with sea water making local food production very difficult. This resulted in the islanders losing their self-sustainability, as they then had to be dependent on a food supply that was shipped to the islands from mainland Bougainville. This also prompted the islanders to think about whether to relocate to Bougainville mainland or stay and perish with the islands. The council of chiefs viewed the dependence on the government and other supporters for food as not healthy or sustainable for the islanders. This led to their taking immediate action to set up Tulele Peisa, which means "sailing the waves on our own" in the local Halia language (Beldi 2016), with the aim of establishing ways to tackle the immediate and long-term impacts of climate change, including finding land to relocate the people.

Tulele Peisa Incorporated, is a national not-for-profit non-governmental organization set up in 2006 in the Autonomous Region of Bougainville by the Carteret Islanders[1] for the well-being of the present and future generations of the Carteret Islands. This can be seen through their explicitly stated goal "to maintain our cultural identity and live sustainably wherever we are" (Rakova 2012: 12). Tulele Peisa was established at the time other key institutions, such as the Autonomous Bougainville Government, were not actively involved and in the absence of core policies such as the Government of PNG's *Carterets Relocation Program* and the *Atolls Integrated Development Policy*. This created some challenges for Tulele Peisa because implementation of their activities was quite difficult without the supporting institutions and policy documents. Nevertheless, Tulele Peisa continued with one of its core projects, which was the *Carterets Islands Integrated Relocation Program* in the Tinputz area of the main island of Bougainville.

Tulele Peisa played a significant role in negotiating relocation processes with the Autonomous Bougainville Government. It also works closely with the Catholic diocese of Bougainville who made available some of their unused church lands in the Tinputz area for the Carterets Relocation Program. At the same time, Tulele Peisa coordinated with the Catholic diocese of Bougainville to relocate the families who were most affected and chose to leave the islands and also ensured they received counseling and similar services from the church. Between 2012 and 2015, Tulele Peisa negotiated with the Catholic diocese to transfer the Tinputz area land title to the families relocated to that land. Tulele Peisa also works with the PNG Border Authority to purchase these lands the families have resettled on after it is surveyed and valued. These purchased lands would be given back to the families.

In April 2009, the first two families were successfully relocated, due to the efforts of Tulele Peisa, to Marau village, located on the west coast, in the South Bougainville District. The area is more than 1.5 meters above sea level. This was followed by the relocation of additional families to Tinputz, which is in the North Bougainville District, about 80 km southeast of the Carteret Islands. These actions taken by Tulele Peisa in collaboration with the people have gone a long way in ensuring that the Carteret Islanders, and their cultural values and practices, are given the opportunity to survive.

Case 2: Community awareness and education— Research & Conservation Foundation PNG

The Research & Conservation Foundation (RCF) is a not-for-profit non-governmental organization established in 1986 and is based in Goroka, Papua New Guinea. RCF works very closely with a range of stakeholders, including customary landowners, in the areas of natural resource management, conservation education to promote biodiversity conservation efforts, and the extension of sustainable rural livelihoods. Through its various programs and activities, RCF recognized the problems that threatened ecosystems, such as prolonged drought and continuous rains leading to flooding and landslides. This greatly affects many rural communities, and the biodiversity within these areas. In the central highlands of PNG, RCF took measures to strengthen the capacity of vulnerable communities to act to minimize the impacts of climate change.

The actions taken by RCF were twofold. The first action involved sharing through teacher training, with elementary and middle-school teachers, basic knowledge about the causes and effects of climate change, and what actions they can take to minimize those effects. This process required working with teachers to develop climate change related resource books, and then developing special training for teachers on how to effectively utilize the lessons and activities in these books. As teachers are involved throughout the process of developing this material, they are key in suggesting the topics and themes that are appropriate for Papua New Guinean children between the ages of 6 to 14. The resource book for children between 6 to 9 years contains basic climate-related topics that look at useful resources, natural and human events, changes in the environment, and taking actions. For example, under the theme useful resources, children identify useful resources in the environment and are introduced to water as a useful resource. They then learn about the effects of having less water during prolonged dry periods such as the soil dries up and becomes hard so food cannot grow well. Or, the creek that supplies freshwater dries up so people have to look further to find drinking water. Those for children between 12 to 14 include topics such as traditional knowledge of weather and seasons, understanding basic science of climate change and its causes and effects, recognizing the serious effects of climate change in terms of prolonged drought and food shortage, recognizing the importance of water as needed for saving life, and identifying various approaches for communities to adapt to climate change.

The use of these teacher resource books have enabled teachers to teach more complicated concepts such as greenhouse effects, the relationships between climate change and global warming, and their possible impact on elementary and primary school children. Teachers are also well-equipped to teach about what the children and their communities can do to minimize the impact of climate change.

The second action on the part of RCF involves empowering communities through community awareness. This approach includes delivery techniques such as public awareness talks on the significance of recognizing impacts of climate change and what communities can do, comedy shows and plays on the causes of climate change and what actions communities can take to minimize these, drama performances by children and community members, community forums, and other formal training. For example, in a play showcasing the effect of climate change, two friends appear in the scene discussing how the sun is too hot and the soil appears to dry up causing their food crops to not grow well. They discuss what might be happening. They are also troubled that someone might be jealous of their skills in growing very big sweet potatoes so they might have put some magic spell on their gardens. The son of one of the two friends appears on the scene and hears their conversation. He then takes the liberty to explain the reason for the increase in the heat, which is drying up the soil so food can't grow well. He tells them it is because of a phenomenon called climate change and he goes on in a simple way to explain its causes and effects and what people can do to prevent that. The two friends ask him a lot of questions and he happily responds with simple answers to ease their worries. They thank him for helping them to understand this problem of climate change and the scene ends there.

All of these approaches are aimed at highlighting the causes and effects of climate change and what actions people can take to address or minimize the impact of climate change. These gatherings also provide people with the opportunity to express their views and opinions and talk about what actions they can take to minimize this. One example of actions taken by communities in response to these awareness-raising programs and training has been to replant trees in areas of anthropogenic grasslands, along hillsides, and near river banks. The trees act as carbon sinks and their roots also hold the soil together to prevent erosion in unstable areas.

Case 3: Scientific research and capacity building— Wildlife Conservation Society PNG Program

The Wildlife Conservation Society (WCS), established in the 1800s as the New York Zoological Society, is an international not-for-profit non-governmental organization with headquarters in the US, at the Bronx Zoo in New York City. WCS focuses on four core global challenges—emerging wildlife diseases, climate change, the increasing harvest of natural resources by local communities, and the expansion of extractive industries. Their mission is to save wildlife and wild places, and support local human interests by fostering sustainable livelihoods to achieve long-term conservation.

The WCS PNG Country program has been involved with conservation efforts in Papua New Guinea since early 1980s. In the late 1990s and early 2000, WCS recognized the threats to wildlife, wild places, and human livelihoods caused by the changing effects of the climate and has taken an active role in researching, planning, and implementing programs to minimize these negative effects in Papua New Guinea. Their efforts include:

- Implementation of sustainable forest management under a Village Reducing Emissions from Deforestation and Forest Degradation Plus (or REDD+) approach whereby a mangrove protection and planting scheme was established. Mangroves are an important plant as they act as an anchor to protect the coastlines and also as carbon sinks.
- Carrying out scientific research on climate change that focuses on developing an adaptive capacity model that ensures climate resilience in communities. In addition, using spatial databases for identifying threats and vulnerabilities, WCS measures impacts of climate change. It also implements different approaches and strategies to improve food security, particularly in terms of planting of appropriate food.
- Capacity building for climate change. This is aimed at developing capacity within the PNG Office of Climate Change and Development, the sub-national government, and local communities, and also empowers people by teaching them to monitor biological and socio-economic change.
- Supporting climate change awareness and education through documenting local knowledge of climate change and adaptation, developing school curricula and community awareness materials about climate change.

The testing out of the adaptive capacity model developed by WCS, as well as how climate change resilience can be created, has been used to project possible scenarios of sea level rise, and their impact on low-lying island communities. In addition, WCS input to the PNG Office of Climate Change and Development has not only enhanced the PNG officers' and civil servants' capacity but also provided technical input that has informed climate change policy development.

Implications for climate actions

The experiences from Tulele Peisa in the Carteret Islands indicate that a community's effort to take ownership of climate-related problems and develop their own adaptive actions, supported by the state and non-state organizations, such as the Church, is a sustainable measure. RCF's experiences, in Papua New Guinea, indicate that use of various approaches to disseminate awareness and education messages on climate change and its impact is an effective method to enable wider members of the community to improve their knowledge and understanding of climate change. Similarly, the experiences of WCS, also in Papua New Guinea, suggest that efforts to carry out research into climate change by empowering civil servants with climate change knowledge not only enhances their capacity but also provides opportunities to develop adaptive measures for the region. These adaptive measures are informed by research and also directly benefit the wider communities. Generally, a common element within the work of these three organizations is that their activities are community driven. Thus, communities are able to collaborate and cooperate in ensuring that the goals of project activities are implemented and adhered to.

These civil society organizations also acknowledge the significance of local or traditional ecological knowledges (TEK), as may be noted in the type of content and information they gather in their projects. The question of whether and in what way TEK is important in responding to climate change needs to be investigated further but other scholarship has shown that TEK does have "the potential to play a central role in both indigenous and nonindigenous climate change initiatives" (Vinyeta and Lynn 2013: i). For civil society organizations in PNG, the inclusion of TEK would be significant because many of the rural communities "still depend on TEK as a form of sustainable resource management" (Tiu 2016: 276). I suggest that an awareness of the environmental context based on TEK is crucial as a key foundation on which to build new knowledge and information about climate change.

Clearly, communities in PNG recognize their predicaments and are willing to take actions to address them. However, full-blown solutions would require substantial technical and financial support to both tackle the issues and/or reduce the negative impacts. It is for this reason that efforts to strengthen the capacity of communities to address climate justice have to be part of a coordinated effort by all stakeholders.

In this context, I offer three suggestions for a sustainable climate future for PNG: research into climate-related issues should be informed by policy, and provide suggestions for actions that are community oriented, and globally acceptable; funding from external sources for climate change should be aligned with community and organizational actions on the ground to be most effective; and a long-term sustainable approach to address these issues should coordinate effort at sub-national and national government levels.

Acknowledgments

I am grateful for the information provided by the following to ensure the work of civil society organizations in Papua New Guinea was shared with others: Ursula Rakova, the Director of Tulele Peisa Inc.;

Dr. Richard Cuthbert, former Director, and Arison Ariafa of Wildlife Conservation Society PNG Country Program; and the Management and Staff of Research and Conservation Foundation PNG.

Note

1. The Carteret Islands are the outer islands of Autonomous Region of Bougainville in Papua New Guinea. The Carteret Islands consist of several atolls and low-lying islands that are highly threatened by sea level rise.

20 | CULTURAL RESILIENCE AND CLIMATE CHANGE: EVERYDAY LIVES IN NIUE

Jess Pasisi

The first time I went to my Dad's land of birth was in 2015. I was 24 years old, had lived with my parents and siblings in New Zealand for all my life, and had just started a doctoral study on displacement. I was scoping out Niue as a potential focus for my research. I decided to bring my parents and my partner along for a bit of a holiday. My Dad hadn't been back to his homeland for 40 years. I remember flying over Niue as we circled to land. It seemed so small, so alone, a 260 km² piece of green in a shimmering sea of blue. Niue is one of the world's largest coral atolls. The highest point on the island is 68 m (223 ft) which is near Mutalau, the village my Dad is from. He remembers having to walk up the hill if he missed the school bus. It always seemed like such a long climb up but, now when he sees it, he thinks it is barely a hill. The majority of the island is covered by regenerating forests and agriculture areas. There are also some high forest and coastal-forest areas that are home to many Uga, a large land crab in Niue that is popular for hunting. The water that surrounds Niue is the lifeblood of its people and has been for their entire history. Like many other Pacific nations, people's livelihoods often depend on the ocean. Yet growing reliance on external imports are having an impact on how younger generations connect with traditional Niuean practices.

It takes just over three hours (2,480 km) to get from Auckland International Airport to Hanan International Airport, Niue. Air New Zealand is the only air carrier to Niue and, until recently, it only flew out to the island once a week (it now flies every Tuesday and Saturday). Since Air New Zealand began its service to Niue in 2005, annual visitor numbers have continually increased from just under 2,800 to over 8,000 visitors in 2015. On average, tourists make up over 60 percent of arrivals in Niue. To attract tourists to Niue's pristine waters, fishing, diving, whale encounters, and its unique landscape are all promoted as an experience of something different to the white sandy beaches of Niue's nearby tropical neighbors. Niueans are also well-known for their friendliness—every car you pass will have someone waving to you, even if you are driving too, something I got used to on the island.

Niueans haven't always been known for their friendly demeanor. Captain Cook referred to Niue as "Savage Island" after attempts to land on the island were met with what he considered to be undue hostility. Niuean accounts perceive the event differently, saying that a fair challenge was made by the Niueans to Cook although Cook and his men used guns to drive the Niueans back into the forest. Failed attempts by missionaries to land on the island in the early days entrenched the "Savage Island" label. Yet, there are accounts of Niueans helping a deserter of a whaling ship find his passage onto another ship as well as reports from John Erskine who visited the island in 1849, referring to Niueans' friendliness, morality, and intelligence (Ryan 1984).

Plane day always stirs up a little excitement in Niue. It's a time to farewell and a time to welcome. Family members reunite, tourists step out onto the tarmac and begin their holiday in a tropical paradise. For me, my holiday would have to wait. After we landed at Hanan International Airport I went straight into a meeting for Tofia Niue, a not-for-profit organization that was to be launched with the aim of conserving and sustainably managing Niue's waters. In Niuean, *tofia* means sea, but it also implies a healthy respect for the sea. The organization was set up as a collaboration between the Niue government, civil society, and the private sector, as part of the Niue Ocean Wide (NOW) project that aimed to conserve and sustainably manage Niue's $316,584$ km^2 exclusive economic zone, which includes Beveridge Reef, a submerged atoll over 200 km from Niue. Some members from the private sector included fishing charter owners, restaurant owners, and land and sea tourism operators. There were also vanilla farmers, whale protection activists, members of the Church, and a varied group of curious people from the community. The government was represented by officials from the departments of agriculture and fisheries, ocean and marine, immigration, and statistics.

Three things struck me about that meeting. First, the meeting, though mostly carried out in English, was steeped in Niuean customs. There was an opening prayer in Niuean by an elder, an open discussion with somewhat equal opportunity for all present to speak, a prayer in Niuean to conclude the conversations, and a large spread of food for everyone. Second, there was an expression of concern by both private sector representatives and community members about the level of control exercised by the government through the structuring of the organization. They questioned how this was to be a collaborative partnership if the people making the final decisions were all linked to government—the organization map showed that various levels of both input and accountability pointed to key people directly linked to or associated with government departments. Third was the level of concern people had for the care of marine life, not just in the sense of fishing, but also with regard to the protection of whales and dolphins whose numbers were declining. Elder people had noticed the changes in the number of whales coming in during breeding months. The whales were no longer as visible from land as they were staying further out from the island. If this organization was about conserving and sustainably managing Niue's marine resources, did that mean care for both humans and non-humans?

Tofia Niue had a guaranteed funding of NZ\$1 million from Oceans 5, a group of philanthropists providing grants to organizations that support ocean conservation. When I asked some of the people present what they thought about the newly formed organization, many were skeptical about what it could do to support conservation of marine life. One fisherman noted that the catch had been decreasing for many years and he hoped this venture might be able to find some solutions. Many were positive in wishing Tofia well, some even put their hand up to fill any positions that might be going—the incentive of a million dollars was not to be sniffed at. But going by their past experiences, some noted that NGOs often talked a good game but seldom fulfilled their promises.

On my second day, I traveled to Tāoga Niue and the national museum (which is next door), both located inland near the island's education institutes, which were recently relocated and rebuilt, following the 2004 Cyclone Heta, with funding from the Australian government. *Tāoga* means treasure, often referring to treasured possessions or belongings. The museum's location is temporary and much of their exhibition is cramped into a small room where the entrance fee is NZ\$10. There was a large display about Niuean soldiers—garments, images, and registers with names of those who volunteered to go to war in support roles as part of the Pioneer Battalion of the British Army, the villages they were from, and their ages. The museum also had displays of traditional weapons, *tika* (wooden javelin), fishing gear, weaving, a large contemporary quilt that had a patch for each village, and some Niuean books ranging from poetry to government records. No dictionaries were on sale though—the Niuean dictionary can be difficult to get hold of unless you know the right people such as the person I was heading to meet next.

Next door to the museum is where I met Moira Enetama, Tāoga Niue's director. Her office is filled with paperwork, books, and relics from the old museum. She has five paid staff who have desks in the adjoining room. I asked her about the impacts of climate change and displacement on the island. Although she didn't know of anyone who had left Niue due to climate change she talked about the displacement of a group of families from Tuvalu who came to Niue in the 1990s. Tuvalu is 1,782km from Niue. Leaders from Tuvalu at the time had been lobbying other nations such as Australia and New Zealand to help relocate some Tuvaluan families. The reasons were not clear but there was talk of overpopulation and climate change.

The stories around the families from Tuvalu migrating to Niue are varied. Jane McAdam (2012) notes that the migration was more of a case of overpopulation in Tuvalu that later became intertwined with climate change. The Tuvaluan government had long sought assistance from New Zealand and Australia in providing land for its people but, ultimately, Niue was the only place prepared to accept the migrants. Nine Tuvaluan families were the first to move to Niue and some believe this to have been a strategic move to gain Niuean citizenship and then use Niue's automatic citizenship arrangement with New Zealand to jump across to the developed world. However, most of the families actually remained in Niue, many intermarried with Niueans, and only a few of the families left to go to New Zealand, mainly due to extenuating circumstances such as to receive health care that wasn't available on the island.

Although many Niueans seemed aware of climate change, particularly the changes to weather patterns such as periods of drought, more severe cyclones, and rising sea levels, the people I met in 2015 didn't necessarily see it having a big impact on their lives, and certainly they didn't see it as a reason to leave the island. Yet, the museum alone provided a stark reminder of how changes in climate were taking their toll. The museum lost over 90 percent of its artifacts and its original seaside building in the capital Alofi during the 2004 Cyclone Heta. All I could see at the site were the remnants of the building's concrete foundation.

Growing up, one of the most consistent things I learnt about Niue was its cyclones. If Niue ever made it to the prime-time news in New Zealand, it was because of the deadly cyclones. In 1990, the year I was born, two cyclones hit Niue: Cyclone Ofa in late January to February and Cyclone Sina in late November to early December. At that time, there was a massive power outage, most private water tank supplies were contaminated, public infrastructure, including the island's hospital, was severely damaged, several were badly injured, and there was even a loss of life. In early January 2004, the Category 5 tropical cyclone Heta caused devastating damage as it passed over Niue. I remember my parents calling my aunt and uncle in Niue and hearing my aunt's trembling voice. This one was really bad. There was significant loss of infrastructure, damage to the coral reef surrounding the island, and Alofi had almost been flattened by gale force winds and high seas. The tourism sector, as well as the coconut and fruit-tree plantations, in Niue were just starting to show promising returns before the cyclone hit but now the economy was shattered again, and schools, businesses, the hospital, and the museum had to be moved inland. A mother and her young toddler were killed.

Although Niue is not located in any traditional geological eye of the storm, the severity of natural disasters that have hit the atoll island has increased over the last 60 years. Cyclones are followed by irregular weather patterns, including long periods of drought and floods, all of which have had devastating impacts on food crops and plantations. One industry particularly impacted by this has been vanilla farming. In the last few days of my trip I spoke to one vanilla farmer who had grown up on the island, moved to Australia and New Zealand for a few years, and returned to run a vanilla farm. She remembered a time when there were nearly 200 vanilla farmers on the island. Today there are barely five left. No matter how many times Niue has diversified to try to build up their trade and export industries, it has been repeatedly brought down by the ravages of a changing climate. Plantations of *talo* (taro), banana, *loku* (papaya), and more recently, *noni* (a medicinal fruit), are dotted around the island but it really is only the hardy root crops that survive. Still, large parts of the island have been cleared of the dense bush for commercial agriculture with much of the cleared sites looking like desolate war zones.

One thing many tourists also notice about Niue is the number of houses that are no longer lived in or cared for. Overgrown vines creep both inside and outside houses almost as if the forest was reclaiming them to retaliate against the fires and bulldozers that were destroying the forest elsewhere on the island. The government has tried to demolish houses that were no longer lived in but locals and Niueans living overseas have resisted because these homes have cultural value. Clearly, place attachment is dear to the heart of Niueans, local and abroad, and pleasing the temporary tourist with aesthetics is not yet more important than saving a family home.

Cultural identity is important to many Niueans but less than one-third of the Niuean population in New Zealand can speak the language. Close to 24,000 Niueans live in New Zealand. It is the largest population of Niueans in the world. While generations of Niueans made the move to New Zealand in the 1960s, '70s, and '80s, there has been some noticeable interest recently from Niueans abroad to return to the island. While I was at the Alofi wharf one afternoon I met a family friend who had been living in New Zealand but had recently decided to move back to Niue. She said that New Zealand wasn't all it was made out to be: "you could wave goodbye to money as you paid for rent for a house you would never own—why not just come back home to Niue where you already have a house?" Tourism was on the rise and one could easily pick up a job or even start a business. Even one of my cousins on the island was looking at starting up a café near the seaside.

Funding from New Zealand, which supports these kinds of sentiments, has made it a priority to support tourism and economic development in Niue. This funding provides incentives to both Niueans and foreigners to start a tourism-related business. No wonder then, there is more development along the island's sea front. The recent history of cyclones and severe weather events hasn't stopped the number of cafés, holiday homes, and luxury resorts along the coastlines. Not long ago, the missionaries too had coaxed Niueans to move from their more central island homes and villages to live nearer the sea. While cultural instinct, which has previously kept villages further from the seafront, calls for cautiousness about (and respect for) the dangers of the sea, it is suppressed by the push to secure economic growth, particularly through tourism. Ironic as it is, developmental funding to Niue is actually pushing the island deeper into climate risk rather than helping its inhabitants to adapt to climate change.

On our last night on Niue we had a final dinner with our family. My Dad had taken me Uga hunting early that morning so we took our haul to the family dinner. The spread of food was incredible. If there's one thing Niueans do well, it's food. The staples really haven't changed from when my Dad was growing up – fish, *maniota* (tapioca), *talo*, and of course the seemingly mandatory dish that is *punu povi* (tinned corned beef). The gathering of family and the community around food is ingrained in Niuean identity. I'll never ever say no to *takihi* (a dish of layers of sliced taro and ripe papaya with coconut cream, traditionally wrapped in banana leaves) and baked in an *umu* (an earthen oven). Knowing that there is a Niuean way of doing things—as simple as a special way to catch an Uga or baking *takihi* in an *umu*—is where I see climate change and culture intersect for Niue.

It's funny how cultural traditions like Uga hunting can stay with a person. My Dad had left Niue at 16 and, apart from intermittent conversations with his family in Niue, he had assimilated into life in New Zealand. But as soon as we landed in Niue it was like smoldering twigs catching fire—he was husking coconuts, picking *loku* (papaya) and "island popcorn," holding forth on traditional plant uses, and setting Uga traps. You can take a Niuean out of Niue, but you can't take Niue out of the Niuean.

As our plane departed Niue I realized that I didn't meet any locals who were pessimistic about the future or about climate change; no one was in despair. Some even pitied me for having to return to New Zealand, the land of milk and honey that now didn't seem so sweet. People were noticing changes in weather patterns, and certainly knew about climate change, but weren't necessarily upending their day-to-day lives to "deal" with it. They saw themselves as different from those of the "sinking islands" such as Kiribati and the Solomon Islands. Was it their cultural resilience or were they being hypnotized into a bubble of tourism-led economic development by foreign aid agencies? That is a question my research seeks to answer.

21 | THE CONTRADICTORY DEVELOPMENT POLICIES OF THE MALDIVES IN THE FACE OF CLIMATE CHANGE

Mohamed Hamdhaan Zuhair

Introduction

My country, the Maldives, is an island nation comprising 25 coral atolls, oriented north to south and located southwest of Sri Lanka. These atolls consist of 1,190 coral islands of which 358 islands are being utilized for human settlement and tourism activities (Ministry of Environment Energy and Water 2007). The average height of the islands in the Maldives is 1.5 m above Mean Sea Level (MSL) (Shaig 2006). The low-lying nature of the Maldives makes the islands extremely vulnerable to climate change and sea level rise. According to the Intergovernmental Panel on Climate Change (Mimura et al. 2007), Small Island Developing States (SIDS) like the Maldives are the most vulnerable and least equipped to adapt to climate change impacts. Such impacts include loss of already scarce land area due to erosion, salt water intrusion into groundwater reducing groundwater quality, impacts on coral reefs due to sea level rise and thermal stress, and major losses to the economy due to loss in revenue from nature-based tourism. Human activities have exacerbated the serious vulnerabilities of this tiny island nation. For example, even from historic times, the reefs, which are the first and the most important line of defense for these islands, have been negatively impacted through coral and sand mining activities. The coral aggregates were used extensively for construction, prior to the banning of coral mining in 1990 (Hameed and Ali n.d.). Sand mining still continues, with sand being the fundamental component for construction, especially outside the capital region.[1] At present, the mining is restricted to a few locations that have to be approved by the government's Environmental Protection Agency (EPA).

At the time of writing this chapter in 2017, infrastructure development projects, like harbor development and reclamation, have had environmental impacts that are larger in scale and greater in intensity compared to the stresses the reefs have experienced historically. Geomorphological studies in the Maldives and similar nations in the Pacific indicate that the islands, especially those not disturbed by human settlement, have adapted well to local scale sea level rise (Kench and Brander 2006; Kench and Ford 2015; Webb and Kench 2010). For example, a study of 127 sparsely inhabited islands in the Marshall Islands showed that, despite localized sea level rise, the total land area of the islands increased from 9.09 km² to 9.46 km² (3.51 mi² to 3.65 mi²) between World War II and 2010 (Kench and Ford 2015). However, the impacts on the reefs and modification of the coasts means that this natural adaptive capacity of these islands is significantly reduced. Thus the urbanized inhabited islands in countries such as the Maldives are unlikely to fare well in the face of climate change.

The Maldives in the international climate arena: a brief history

Given the size and vulnerability of the Maldives, it is instructive to note the country has always been quite vocal in the international arena regarding climate change. Even before the democratization of the country, the autocratic leader who ruled the country for 30 years, President Maumoon Abdul Gayoom, had voiced serious concerns about the possible impacts of climate change and the corresponding rise in sea level. Through the leadership of the Maldives, the leaders of SIDS met in Male', the capital of the Maldives, in November 1989 (Lewis 1990). *A Male' Declaration on Sea Level Rise and Global Warming* was signed by the leaders (Lewis 1990), which was instrumental in establishing the Alliance of Small Island States (AOSIS). Established in 1990, the Alliance brings together the most vulnerable group to climate change impacts, the SIDS. It consists of 39 members and five observers (Climate Policy Observer n.d.). The aim of the Alliance is to provide the unanimous voice of SIDS to the international climate arena. In this regard, President Gayoom's leadership was instrumental in establishing one of the most dominant voices on climate change in the international arena (Climate Policy Observer n.d.).

Democratization of the Maldives occurred in 2008 with the ratification of the new constitution (Henderson 2008). The executive, judicial, and legislative branches of the government were separated and the absolute power of the president reduced (Henderson 2008). The charismatic leader, President Mohamed Nasheed, won the first multiparty presidential election held in 2008 (Henderson 2008). Like his autocratic predecessor, President Mohamed Nasheed was very vocal on climate change and environmental justice. Considered by many as the democratic champion of the Maldives, combined with his charismatic nature, meant that Nasheed was an immediate hit in the climate arena. He was regarded as one of the stars at the Conference of the Parties (COP) in Copenhagen in 2009. Bold moves by the president, like pledging to make the Maldives carbon neutral by 2020 and a first-of-its-kind cabinet meeting underwater increased the president's popularity abroad (Duncan 2009; Ramesh 2009). Moreover, he pledged to start a fund to buy land from abroad so that Maldivians could migrate there, as needed. Within the academic literature, many regard these efforts as mostly symbolic, as the targets he set were unrealistic and unlikely to be achieved (Kelman 2014; Kothari 2014). Despite this, his bold statements put the Maldives at the forefront of international climate headlines. However, his presidency was short and he was forced to resign in February 2012 through what many now regard as a "silent coup." His then vice president, President Dr. Mohamed Waheed Hassan, was sworn in as the president, completing his term in 2013.

The highly contentious 2013 presidential election was won by President Yameen Abdul Gayoom. Following his election, the Maldives reverted back to a form of autocratic rule, as the president took back full control of the judicial and legislative branches of the government. Economic development was the prime focus of the presidential campaign; thus, it is no surprise that environment is not the top priority for him in terms of domestic issues. Despite the domestic policy direction, internationally, the Maldives continued to play a prominent role in the climate change arena, with the Maldives becoming the chair of AOSIS in 2015 (Ministry of Environment and Energy 2015). The international stance of the Maldives is reflective in the statement released by the then president following the withdrawal of the US from the Paris Agreement. The statement reads:

> The Maldives, Chair of the Alliance of Small Island States (AOSIS), regrets the decision of the United States to withdraw from the Paris Agreement. Not only does it represent a blow to the global efforts against climate change, it is also a setback for the multilateral process that is essential for governing the international community. (The President's Office 2017)

Government policies and climate change

The highly dispersed nature of the Maldives means that basic infrastructure development is required in all the islands. To provide these services to all the islands in the archipelago requires significant funds. Thus, the Maldives is highly reliant on foreign aid and loans to meet this demand. For example, as of the 2017 budget, 45.3 percent of the funds for public sector investment projects came from foreign aid and loans (Ministry of Finance and Treasury 2017). The vulnerability of the Maldives to climate change is a prime selling point in attaining foreign aid and loans. Hence, regardless of the domestic political situation, the Maldives will continue to rely on these climate funds.

Given that the government came in with an economic agenda, the policies of the government are geared towards economic prosperity, sometimes at the expense of and by selling the environment. As the three pillars of the government are under the complete control of the president, bringing about these changes were straightforward. The constitution has been amended, such that foreign parties can now own 70 percent of the land that they reclaim (Mohamed, Naish, and Rasheed 2015). Lagoons have been leased long term for tourism development, where artificial islands are created through reclamation projects (Nafiz 2017a; Nafiz 2017b). Sedimentation from such reclamation projects affects the coral reefs primarily by blocking off sunlight. This impacts the algae zooxanthellae that live within the coral tissue and upon which, through photosynthesis, the corals rely for energy (Erftemeijer, Riegl, Hoeksema, and Todd 2012). This leads to reduced growth and much greater mortality of corals (Erftemeijer, Riegl, Hoeksema, and Todd 2012). As highlighted previously, coral reefs are the prime natural defense that the island nation of the Maldives has against climate change.

Moreover, the government has taken steps to reduce the power and effectiveness of the Environmental Impact Assessment (EIA) system of the Maldives, which is the only check against environmental damage from development projects. In this regard, through amendments made to the Tourism Act, the regulatory function of the EIA for tourism developments has been transferred from the Environmental Protection Agency to the Ministry of Tourism, a ministry whose mandate is primarily to ensure rapid tourism development and expansion (Zuhair and Kurian 2016). Moreover, the EIA regulations have been amended, such that the period of review of EIA reports have been shortened, thus reducing the thoroughness of the review process (Zuhair and Kurian 2016).

All of the above creates a situation where the country continues to rely on international donor funds for climate change adaptation projects, which are protective of the environment, while, simultaneously, undertaking infrastructure development through government funds, private investments and foreign assistance not affiliated to climate change, which are, usually, destructive in nature because they are undertaken without proper environmental considerations.

Development trends in the Maldives

Looking at some planned and recently implemented projects will reveal the contradictory nature of developments taking place in the Maldives. The Climate Change Adaptation Project (CCAP)—a European Union-funded, World Bank-managed project—has been undertaken primarily in two southernmost islands, Hithadhoo and Fuvahmulah (World Bank 2015). These two islands have some of the most significant wetlands found in the Maldives. One of the key goals of the project is to restore and establish a well-managed Protected Area system within these wetlands (World Bank 2015). The project aims to achieve this by opening up the area for ecotourism through comparatively simple structures such as boardwalks and visitor centers, thus giving an economic incentive for protection (World Bank 2015). The mangrove vegetation found in association with such wetlands plays a significant role in coastal protection and hence is a key natural climate defense, along with the coral reefs. The experience of the 2004 tsunami suggests that mangroves played a key role in absorbing and weakening wave energy. Thus, islands with mangrove vegetation fared better in that tsunami (Bluepeace 2007). At the other extreme, a contract has been awarded and surveys are underway for a project that involves the development of an airport in one of the northernmost islands of the Maldives, Kulhudhuhfushi; this development requires reclamation of a wetland area (Naish 2017). This, despite the fact that an international airport already exists in Hanimaadhoo island, an island just a 30 minute speedboat ride away.

Another key area for which the Maldives receives climate funds is for coastal protection of eroding islands. For example, the government of the Netherlands has allocated grant money of approximately 20 million euros to undertake coastal protection of the severely eroding northeastern beaches of Fuvahmulah (Saleem 2016). Similarly, in Thinadhoo, another southern island, through Integrating Climate Change Risks into Resilient Island Planning which is a project initiated by the United Nations Development Programme (UNDP) and the Global Environment

Facility (GEF), a 950 m (3,117 ft) geobag revetment[2] was put up on the eroding western beach (Ministry of Environment and Energy 2014).

While these coastal protection projects do protect eroding beaches, at the other extreme, projects that lead to the destruction of coral reefs, the natural barrier that protects the islands, are implemented simultaneously. In the name of "integrated tourism development" many pristine lagoons in the capital region are being reclaimed for tourism development. Taking a leaf out of the book of Dubai, these artificial islands will have yacht marinas, duty free shops, retail outlets, and resorts (Nafiz 2017a, 2017b). One such project in Emboodhoo Falhu, one of the largest stretches of lagoon in South Malé Atoll, is undertaken by the Thai group, Singha Estate Public Company Limited (Nafiz 2017a). The reclamation involves creation of six islets, each of which has its own specific purpose such as resort accommodation and duty free shops. The project is expected to create many jobs within the capital region, where one-third of the population resides. Thus, economically, the project is justifiable on the grounds of domestic development. However, a weakened EIA system, governed by the Ministry of Tourism, means that environmental considerations were largely ignored when approving and starting this project. Aerial imagery of the works undertaken at the site shows significant sedimentation that is likely to destroy vast stretches of the Emboodhoo reef.

The way forward

Some argue that an authoritarian system of governance, as has been practiced in the Maldives, is in fact best placed to undertake climate adaptation (Beeson 2010; Shearman and Smith 2007), because immediate and, at times, controversial actions may be taken at a comparatively fast pace (Beeson 2010; Gilley 2012; Shearman and Smith 2007). However, for this to be effective, political will to put the environment first should be there. Moreover, the experience from countries such as China suggest that quick action taken through a top-down decision-making process, without the involvement of the affected public, is not very effective in terms of compliance (Gilley 2012). The experience from the very vulnerable Maldives indicates that in an economically driven, authoritarian system, environmental sustainability does not become the top priority. Hence, instead of authoritarian environmentalism, authoritarian environmental destruction is practiced. Yet, the dispersed nature of the Maldives and the huge capital cost required for development means that from an economic perspective, it makes sense for the country to continue to support the climate movement internationally, and obtain climate funds for development. In a way, the need to rely on external funds is a positive, as without such funds, environmentally centered developments are unlikely to occur. However, from the perspective of holistic environmental sustainability, the policies of the government mean that the damage done in the name of economic development is likely to negate any positive outcomes from such projects.

This is not to say that a representative democracy in the Maldives will necessarily fare better in the face of climate change. Under the wrong leadership a representative democracy is also bound to fail in tackling climate change, as exemplified by what is happening in the US currently under President Trump. The US apart, Justin Trudeau, the Prime Minister of Canada, demonstrates that even the most liberal leaders can take controversial decisions within a representative democracy, as he has approved additional pipelines to carry a million barrels of tar sands crude oil per day from Alberta (McKibben 2017). At the time of writing this chapter in 2017, the Maldives government put forward an amendment in parliament to reduce duty on petrol and diesel from 10 percent to 5 percent (Nasih 2017). A party led by a climate champion should logically oppose such an amendment. However, the opposition party led by former president Nasheed, went one step further and proposed that petrol and diesel should be totally exempt from any duty (Nasih 2017). While this might be simply politics in play, this approach shows that to gain popularity, even the most progressive of leaders take controversial actions. In other words, the reality of a political climate under a capitalistic representative democratic system of governance is such that politicians too often succumb to populist measures. Politicians in such a system are also beholden to those who fund their political campaigns, such as big businesses, whose interests are often in direct contradiction to environment protection.

In terms of environmental performance world-wide, consensual democracies, such as Germany and Switzerland, fare the best (Dryzek and Stevenson 2011). In such a system, informed deliberations lead to consensus on issues (Baber and Bartlett 2005; Dryzek and Stevenson 2011). Thus, the validity of an argument is decided based not on popularity but rather, reason and logic (Baber and Bartlett 2005; Dryzek 2009; Miller 1992). Perhaps, when a more consensual system of governance is established in the Maldives, where decisions are made based on informed deliberations, decisions that weaken the climate defense of this fragile island nation will not be taken. In such a system, it is unlikely, for example, that amendments to reduce the power of the EPA, or to reduce duty on petrol and diesel will be passed. As of 2017, the Maldives is very far from establishing such a system, as the government does not accommodate a plurality of voices on such policy issues.

Even though there is not much that can be done domestically, perhaps internationally, there is much to aim for. Similar to the pressures heaped upon the developed world to reduce carbon emissions, pressure needs to be applied on developing countries to protect the natural defenses that these countries have against the negative impacts of climate change. One way to apply this pressure is to insist that developing countries hold themselves accountable to conditions and certain commitments for the award of climate funds. After all, the donors hold a lot of cards, and countries such as the Maldives are very much dependent on foreign aid and loans. More fundamentally, it is only when the destructive logic of the current global fossil fuel-driven economic and political system shifts to one that is holistically sustainable that we can expect green democratic development in the Maldives. Unlike some other parts of the world, however, the Maldives cannot wait too long for such a climate-friendly future to emerge.

Notes

1. The centralized nature of the Maldives means that there is wide disparity in terms of resource availability and affluence. In the central capital region, imported river sand is widely available and used for construction. In the outer atolls, the communities still depend on cheaper locally mined sand.

2. A geobag revetment is a seawall made of geotextile bags filled with sand.

22 | WHY CAPACITY BUILDING NEEDS TO DO JUSTICE TO THE GLOBAL SOUTH: INSIGHTS FROM BANGLADESH

Naznin Nasir, Meraz Mostafa, M. Feisal Rahman, and Saleemul Huq

Introduction

There was a heated discussion at the December 2015 COP 21 in Paris between the Global North and South around a topic that has historically been uncontested—capacity building (Huq 2016). Developing countries argued that the existing capacity-building programs were ineffective and needed a major overhaul because they were not of much use to nations and communities most vulnerable to impacts of climate change. The developed countries, on the other hand, were content with their current approach towards the issue. After considerable discussion, Article 11 of the *Paris Agreement* (UNFCCC 2015) reaffirmed that capacity building and climate education are essential to climate action, and a separate decision document included setting up a committee on capacity building.

While capacity building has generally been understood as a managerial issue to make climate action in developing countries more effective, we argue that it is also just as much a climate justice issue. Let us begin with an overview on capacity building in the context of climate change. The term capacity building has a range of definitions and is used slightly differently in different contexts. Capacity has been suggested to be a combination of attributes which allows people to act together to take control over their own lives in some fashion (Morgan 2006). Within the United Nations Framework Convention on Climate Change (UNFCCC), capacity building is defined as "the processes of developing technical skills and institutional capability in developing countries and economies in transition to enable them to address effectively the causes and results of climate change" (UNFCCC 2014). Capacity building can thus refer to workshops or training that can help government workers access international climate finance more easily; it can refer to proposal-writing programs that help institutions in developing countries access various project funding opportunities; it can be about developing better monitoring and evaluation systems. While it tends to refer to building the capacity at the institutional level, it can also refer to programs that help communities on the ground learn skills to be better able to adapt to the impacts of climate change.

Perceiving capacity building in the context of climate change as an apolitical issue has some underlying implications as it gives a partial understanding of the reasons behind the failure of capacity-building initiatives in addressing the capacity gap of the developing nations. The failure has to do with more than just ineffective strategies. There are deep-rooted political and economic reasons as well. The failure of capacity-building initiatives affects those that are already marginalized because of the historically uneven development between the North and South. Developed countries were able to grow their economies and build their capacities through the burning of fossil fuels. Developing countries on the other hand have contributed very little to climate change, having used only a fraction of fossil fuels consumed by the developed world. The "uneven development" within and among countries has translated into people in different places having unequal access to wealth, skills, and resources. In the context of climate change, this "uneven development" has meant that some people are better placed to mitigate and adapt to climate change than others. Vulnerability to climate change impacts, among many factors, is also interrelated with the capacity of people to build resilience. Not appropriately addressing capacity gap issues, therefore, means that the vulnerable continue to remain vulnerable. This is why capacity building of different stakeholder groups to implement resilience-building measures is necessary to achieve justice (Bapiste and Rhiney 2016).

Capacity building unfortunately remains a small part of worldwide climate action, particularly when it comes to the area of adaptation. The existing model for climate change capacity building emerges from capacity building in development at large. However, given that international climate action is premised on notions of justice, capacity building within the context of climate change should be significantly different from that of development. A just capacity-building program will benefit the developing country more than the donor country. Although headway has been made at the Paris UN climate talks to address the need to create a more efficient and fair capacity-building system, there is still a long way to go to actually implement it. We highlight some gaps within existing climate capacity-building programs and also recommend means to address the identified gaps. We also argue for the creation of "leave-behind in-country" capacity systems and capacity suppliers.

Capacity building: history of ineffectiveness and inefficiency

Despite a wide range of efforts, both bilateral and multilateral, most developing countries still face substantial capacity gaps which hinder the climate actions they intend to pursue. The inefficiency and ineffectiveness in capacity-building initiatives continue to persist mostly because of short-lived ad hoc project-based interventions, lack of investment, and lack of country ownership (Huq 2016; Khan, Sagar, Huq, and Thiam 2016; VanDeveer and Dabelko 2001).

Since the notion of development cooperation by industrial countries began in the 1950s in the form of technical assistance to developing countries, strategies for ensuring aid effectiveness have kept changing. The focus of development cooperation has shifted from "aid effectiveness" to "development effectiveness" (Khan et al. 2016; Mawdsley, Savage, and Kim 2014). The emergence of new donors from both the Global North and South and new stakeholders including stronger civil society participation is changing the landscape of development cooperation. As a result, numerous new projects have mushroomed but there is not much evidence of their effectiveness (Keijzer 2013; Keijzer and Janus 2014).

As we have seen first-hand in the case of Bangladesh—one of the least developed countries in the world (GDP per capita US$1,602 as of May 2017) and also one of the most climate-vulnerable nations, development agencies tend to assign a consultancy company from their own country on an ad hoc basis to a developing nation. Such a company then sends its consultants to the assigned nation to hold short-term workshops for various stakeholders. These workshops usually last a few days and are fairly inefficient because the consultant flown in does not know the local language. Typically these workshops involve presentations and participants are deluged with information. After a few more visits by the consultants and a final report, the agencies claim that capacity building has been attained (Huq and Nasir 2016). The workshops fail to build long-term capacity. Indeed, as some suggest, such donor-driven, short-term, ad hoc exercises by foreign experts often end up harming local capacity building as these weaken local ownership and do not allow local staff to take responsibility for the project (Godfrey et al. 2002). This has led to a situation where there is a serious lack of capacity within various institutions and government units to address climate change issues.

For instance, a recent report titled "Capacity Building Strategy for Climate Mainstreaming"—put together by the Planning Commission of Bangladesh—found that in attempting to mainstream poverty, environment, climate change, and disaster issues into the Planning Commission, only 10 percent of planning professionals were able to do an adequate job. As many as 30 percent performed poorly while the other 60 percent were able to do so only moderately (General Economics Division 2014). This shows a serious lack of capacity within one of the most important government agencies in Bangladesh: the Planning Commission. Similarly, Ahmed, Huq, Nasreen, and Hassan (2015), in their report on climate change and disaster management sectoral inputs towards the formulation of the Seventh Five Year Plan (2016–2021), also mentioned that there is a deficiency in capacity and managerial skills which has hindered implementation of national climate action plans in Bangladesh. In sum, relying too much on foreign consultants has been a major barrier for Bangladeshi organizations to achieve self-sufficiency in addressing climate change issues.

What is even more of a concern is that the money earmarked for climate capacity-building initiatives in developing countries, by and large, goes to the donor countries' agencies. There is in fact hardly any research on how much money is going towards climate capacity-building initiatives. Nakhooda (2015), analyzing multiple climate funds, argued that despite having many initiatives, the sum of finances provided towards capacity building are rather modest, and the support is rarely enduring and sustained. According to her, capacity-building programs were "useful to the individuals and their anchor institutions, but wider impact on the system [was] often less clear" (Nakhooda 2015: slide 7).

Capacity-building programs are preferred as an umbrella concept by both funding and implementing entities so as to package and legitimize different projects and programs under them (Khan et al. 2016). Given the absence of any substantive data about capacity-building funding, we use total funding directed towards adaptation and mitigation to indicate the disparity that exists between capacity-building funding for adaptation and mitigation. In general there is a significant gap between adaptation and mitigation funding where more money is being spent to build capacity in the area of climate change mitigation as opposed to adaptation. For example, the Green Climate Fund (GCF), set to become the main fund to implement the *Paris Agreement*, to date has allocated only 27 percent of the approved funding to adaptation as opposed to 41 percent for mitigation (GCF 2017). The funding disparity between adaptation and mitigation initiatives is also apparent from the global climate finance data. The Climate Policy Initiative (Buchner, Mazza, and Falzon 2016) reported that in 2013–2014, globally US$ 336 billion was spent for mitigation as opposed to US$27 billion for adaptation. Huq (2016) notes that while most of the demand from developing countries, and especially from the poorest ones, is capacity building on adapting to climate change, much less money is allocated to capacity building for adaptation. Over the 2009–2015 period, support for mitigation-related capacity building was reported to have increased from US$15.75 million to US$321.16 million (Khan et al. 2016). It is evident, therefore, that the current funding paradigm is designed to meet the priorities of donor nations rather than those of developing nations. The fundamental question remains as to how to grow local ownership and leadership in aided projects on capacity building.

To make matters worse, there is no guarantee that national governments will allocate funding on capacity building. For example, the Bangladesh government created the Bangladesh Climate Change Trust Fund to implement priority climate change programs identified in the Bangladesh Climate Change Strategy and Action Plan. Going by a presentation made at the Bangladesh Climate Change Trust (BCCT), the trust fund as of September 2016 has committed approximately US$260 million to 377 projects, of which only 2 percent has been allocated for capacity building (BCCT 2016).

Post-Paris capacity building: learning from history

Since capacity building in the context of climate change is based on principles of climate justice, a just capacity-building program needs to focus on channeling most of the benefits to developing countries. Paragraph 1 of the *Paris Agreement*'s Article 11 says that

> Capacity-building under this Agreement should enhance the capacity and ability of developing country Parties, particularly countries with the least capacity, such as the Least Developed Countries (LDC), and those that are particularly vulnerable to the adverse effects of climate change, such as Small Island Developing States (SIDS), to take effective climate change action. (UNFCCC 2015: 15)

The Parties also agreed to set up a new Paris Committee on Capacity Building (PCCB) to address gaps and needs, both current and emerging, in implementing capacity building in developing country Parties and further enhancing capacity-building efforts, including with regard to coherence and coordination in capacity-building activities under the Convention. Since then at COP22, the PCCB was formed and the committee had its first convening meeting in May 2017. The *Paris Agreement* also agreed to create a Capacity Building Initiative for Transparency (CBIT) to build the institutional and technical capacity of developing country Parties in meeting the transparency requirements of Article 13. In June 2016, the Global Environment Facility (GEF) council approved the establishment of CBIT Trust Fund and allocated US$50 million to the trust fund. Furthermore, to provide programmatic direction of the CBIT Trust Fund, the council agreed to give priority to projects submitted by countries such as LDCs and SIDS, which are most in need of capacity building for transparency-related actions. The council also decided to fund long-term sustainable mechanisms for transparency-related capacity building that allows for national ownership of capacity-building efforts (Khan et al. 2016). These initiatives indicate that the *Paris Agreement* has set the stage for addressing the gaps within the existing capacity-building programs.

Yet, as discussed earlier, short-term technical assistance-based and consultant-driven projects are unlikely to result in sustained institutional capacity. Although universities are among the most sustainable capacity-building institutions in the world, very little climate capacity-building funding has been awarded to universities. While university systems globally are built to produce new knowledge and act as innovation and learning hubs, and indeed have been involved in "producing, communicating, and learning climate knowledge and skills" (Hoffmeister, Averill, and Huq 2016: 1), Southern universities often have limited capacity. Many of these entities lack research and training facilities, and do not have dedicated teaching programs on climate change. Proper funding support could be useful for Southern universities to improve the quality of education, update curricula, improve research and training facilities, and offer other skills-building training to students. Southern universities are beginning to collaborate with and learn from each other and enhance their climate capacity. As recently as April 2017, ten LDC universities have formed a consortium called the LDCs Universities Consortium on Climate Capacity (LUCCC). The intention of the consortium is to strengthen climate capacity especially on community-based climate change adaptation of the partner institutions "to do research and provide training at national level" (Phakathi 2017). Networks and consortiums such as LUCCC could provide useful guidance to PCCB in achieving their mandate.

Similar platforms outside universities could also be envisaged among other stakeholders, especially in the South, across all scales to allow greater interaction, promote knowledge and experience sharing, and enhance collaboration with a goal to enhance climate capacity (Khan et al. 2016). While we can discuss how capacity-building projects can be implemented, there is absolutely no question about the need to ensure adequate and appropriate funding support towards climate capacity building (Khan et al. 2016). While the *Paris Agreement* has included provisions to address some of the gaps in the existing capacity building, it does not guarantee resources. As Khan et al. (2016: 16) note, "the use of the term 'should' instead of 'shall' in the context of developed countries enhancing support for capacity building indicates a lower form of obligation." Not only is it essential for developed countries to fulfil their obligation, it is also necessary to track capacity-building finance.

In conclusion

Capacity building, a "fundamental precondition" of the post-2020 climate regime, will require equal and active participation from all. There are historic reasons that developing countries lack the capacity to address the impacts of climate change. By this logic, as this chapter has argued, developed countries are obligated to help developing countries build their capacity to adapt to climate change. However, the current paradigm of mindless spending of limited allocations in the field of capacity building on presentations and workshops by foreign consultants is clearly failing to cover the capacity gap. Short-duration, ad hoc, and consultant-driven capacity-building programs together with a lack of investment and country ownership are key reasons for the inefficacy of the existing programs. While current practices need to be redefined, and gaps identified, politicians and development agencies will also have to change the way they think about capacity building to help in the urgent task of climate adaptation in vulnerable, developing countries.

23 | CHINA: CLIMATE JUSTICE WITHOUT A SOCIAL MOVEMENT?

Fengshi Wu

The trajectory of China's participation in global climate politics has been anything but linear. In the late 1980s, China was among the first group of developing countries that paid regular attention to the issue and attended most international negotiations. The Chinese government gradually became an enthusiast of the *Kyoto Protocol* signed in 1997. However, once the country's economy and energy consumption took off dramatically in the early 2000s, and its carbon emissions exceeded the United States around the mid-2000s, China turned into a major target of complaints at international climate negotiation tables. The 2009 COP15 in Copenhagen witnessed a near fall-out between the Chinese delegation (headed by then premier Wen Jiabao) and OECD states. Four years later, history took another turn, and China announced it would set an absolute cap on its carbon emission (starting in 2016) and peak its carbon emission by 2030. At the Paris COP in 2015, China joined the world and signed up for the new mechanism of global carbon governance with specified and voluntary-based Intended Nationally Determined Contributions.

While the Chinese government has become aligned with the mainstream at the global level to fight against climate change, there is still very little progress at home in incorporating a justice narrative into the broad policy agenda on climate change. Many factors, including the ones highlighted in this volume and illustrated by other country studies contribute to this disappointing outcome. Despite the rapid rise of environmental activism in the country, there is a lack of a domestically grown social movement to address climate justice as a whole. China has had a fairly developed environmental civil society with over 20 years of history and a visible network of climate concerned non-governmental organizations (NGOs) but almost none specifically works on domestic climate-related justice issues. Without sufficient bottom-up public participation and coherent social activism, climate politics in China remains primarily driven by the central authorities and centered on energy efficiency and carbon emission reduction. Thus, climate policy-making in China is still weak in recognizing far-reaching and long-term impacts of climate change on local society and social justice beyond inter-state equity.

This chapter starts with a brief summary of the main shifts in the Chinese government's general attitudes on climate change and justice, and then explains the main non-state actors and NGOs in the field and their main contribution to voicing alternative opinions. Juxtaposing both governmental and non-governmental climate initiatives sheds light on the fact that NGOs are late comers and less than junior partners, in contrast with governmental agencies, in China's climate politics. In addition to pointing out this general pattern of political dynamics and the unfinished task of merging climate policy discourse with general discussions on political reform and social justice in the country, at the end of this chapter I call attention to a number of issue areas in China that could present more visible evidence of differentiated vulnerabilities to climate change in future.

Shifts in the Chinese state's stance on climate justice

The Chinese government has always been conscious of the issues of equity and North–South disparity in global governance in general and climate negotiations in particular. Though little known to the public, China played a critical role in the formation of the principle of Common but Differentiated Responsibility at the global level. During the last sessions of the UN Conference on the Human Environment in 1972, China, led by its legendary premier Zhou Enlai at the time, proposed ten points for further discussion and drew a clear line between the culpability and responsibility of developed and developing countries. As a result, two entirely new principles were added to the *Stockholm Declaration*, underlining the need to balance development with environmental protection, and every country's right to decide its own environmental standards (Stalley 2013). This is the beginning of the emergence of the Common but Differentiated Responsibility principle that would become a main doctrine for global climate governance and China's stance on climate justice until not long ago.

It was around 2013–2014 that a major, long-awaited shift of top Chinese decision-makers' mentality on climate issues took place. Before this turn, the overall undertone of China's climate policy was firmly anchored on development. The opening of China's first *State Policy*

Paper on Climate Change, published in 2007, clearly asserted: "Climate change is a most important global issue generally concerned. It is both an environmental and a developmental issue as well. But, ultimately, it is a developmental issue."[1] This guiding idea is reiterated and explicated in the Paper's Section III as "to control greenhouse gas emission and to enhance sustainable development capacity are the main goals; to ensure economic development is the core [of all climate change-related actions]." Holding steadfast to such a view on climate politics, China behaved accordingly in international climate negotiations until the early 2010s (during and after the Copenhagen COP), consistently advocating for developing countries' right of economic development, the significance of developed countries' historic damage to the earth's atmosphere, the principle of differentiated responsibilities to reduce carbon emissions, and the necessary transfer of both financial and technological support from developed countries for both mitigation and adaptation.

Two important state-level policy papers—the first *National Climate Change Adaptation Plan* and a new *State Policy Paper (or National Plan) on Climate Change*—were released by the central government in 2013 and 2014 respectively, and produced much new thinking over climate politics inside China. Not only was the once firm stance that "climate change is ultimately a developmental issue" dropped out, but the new policy documents also elevated the importance of adaptation and greatly expanded the scope of climate-related policy-making. The 2015 annual *Report on Climate Change Policies and Actions* provided more detailed explanations of climate-related vulnerabilities, potential harm and relevant preventive measures in the country, and also for the first time included a section on direct public participation in combating climate change.[2]

In recent years, China has modified its overall stance on climate change and the relevance of climate change to domestic and international politics. There are three main features of the new stance: First, China will bear its responsibilities of greenhouse gas emission reduction, with concrete targets, reforms, and policies.[3] Second, China acknowledges the equally important task of adaptation. The key words used by China's Special Envoy on Climate China commenting on the *Marrakech Action Proclamation for Our Climate and Sustainable Development* were "balanced"—balanced between emission reduction and adaptation, and "reflecting opinions from all sides"—the sides of large developing economies (such as China, India, Brazil), developed countries, and the most climate-vulnerable countries.[4] Last but not least, China wants to take up bigger roles in global climate governance, especially in facilitating

South–South climate cooperation. Increasingly, China's official rhetoric on climate change is becoming integrated with its cross-continental mega proposals of international cooperation such as the *Belt and Road Initiative*[5] and president Xi Jinping's narrative of China's "great renaissance" (Xi 2015). These shifts in rhetoric and normative expression certainly reflect deeper changes in Chinese top leadership and its vision of global politics. Nevertheless, they should not be interpreted as a complete departure from China's traditional foreign policy, particularly its stance on sovereign rights and the principle of "peaceful co-existence," and its self-proclaimed solidarity with all developing countries. It should be expected that China will continue to refer to the principle of "differentiated responsibilities," along with other principles newly embraced by the state, in climate negotiations in the post-Paris era.

Climate activism and scholarship on climate justice in China

Environmental NGOs and environmentalism-minded scholars, to some extent, have spotted what is missing in the mainstream climate change and justice narratives in China, and also made various efforts to modify the narratives in their own ways, such as introducing the full range of discussions on the topic from the international level to the domestic audience, presenting critical views of existing climate practices (and legislation) in China, advocating for climate concerns in China's increasing eco-footprints overseas, and other issues.

In the past three decades, civil society, in spite of all the political constraints, has been growing steadily in China, the environmental activism community is perhaps the most developed sector (Hildebrandt 2013; Wu 2013). Of all of the policy fields, not only does the environmental field have the largest number of specialized NGOs, the longest and most continuous history of NGO development and public campaigns, and the broadest scope of work focus, but also has generated the highest level of international recognition (measured by the number of Chinese environmentalists getting international awards) and policy impact. According to research by the China Environmental Organization Map online, there are at least 1,996 grassroots environmental NGOs across all provinces.[6] These local green groups are much better connected with each other for mutual support than their counterparts in other policy fields and social sectors.

However, of the highly vibrant, networked, and fast expanding green civil society in China, specifically climate-related groups occupy a minor portion. The China Civil Climate Action Network (CCCAN)—the only broad network of non-state actors in the climate change field—has only a dozen NGO members. The first and still the only NGO devoted to youth participation in climate action, China Youth Climate Action Network (CYCAN), started in 2007 as an ad hoc group of young leaders from the Chinese environmental activism community who attended the COP15 meeting as a group for the first time. The work focus of both of these platform NGOs is highly in line with the Chinese state's main policies on climate change, such as public awareness, energy efficiency, clean energy, and raising China's voice in international negotiations.[7] While CYCAN targets college youth, CCCAN focuses on learning opportunities for NGOs, tertiary schools, and teachers. CYCAN's flagship activity is the annual International Youth Summit on Energy and Climate Change, and in 2015 the main topics included "energy revolution in China," "smarter campus," and "how to write articles to track the climate negotiation," all of which fit perfectly with the state's main agenda on climate.

There are a small number of Beijing-based NGOs with a higher level of research capacity that have started working on climate policy advocacy, such as the Global Environmental Institute (GEI) and the Greenovation Hub (GH). GEI has a relatively longer history, bigger budgets, and a larger team of staff. Its founders include veteran environmentalists, who were among the first group of Chinese NGO representatives that started participating in international environmental conventions at Johannesburg in 2002. Since 2004, GEI has pioneered in non-governmental policy advocacy in sustainable forestry, climate financing, and rural energy innovation. GEI succeeded in persuading and working together with the state authorities to formulate a guidebook of sustainable practices for all Chinese wood-processing companies with overseas operations, particularly in the world's most valuable forested regions.

Established in 2012, GH has since grown up fast and its Research Unit (based in Beijing) has produced a number of high quality independent reports on climate change, sustainability, and corporate social responsibility. For example, GH, together with partner NGOs, conducted the first independent assessment of "green financing" (i.e., financing for renewable energy) by major Chinese commercial banks and produced a guideline of responsible renewable financing in 2012. This project was further extended in 2015 to track Chinese banks' overseas financing activities in the fields of mining, forestry, dam construction, and basic infrastructure building. Since 2016, GH has started a new policy project to advocate for mainstreaming sustainability in China's *Belt and Road Initiative*, which intends to enhance logistics routes and land and marine time connectivity among China, Central Asia, Europe, and Southeast Asia. Like GEI, GH has called much needed attention to China's global footprints and the socio-ecological responsibilities of Chinese businesses beyond borders.

In early 2014, GH published the first and probably the only report devoted to the topic of climate justice in the Chinese language by a Chinese NGO.[8] In this report, GH introduced the concepts, theories, and practical principles related to climate justice most commonly used at the international level in English. Unfortunately, the report included barely any evidence, case studies, or discussion of climate (in)justice in the Chinese context. Also, reception of the report has been minimal, even among the environmental activism community.[9]

It is not just the NGOs that have been unable to promote an alternative narrative on climate change and/or climate justice in China. Neither have the scholars. A quick content analysis of all the academic publications available at the Chinese National Knowledge Infrastructure (CNKI) reveals the main patterns of scholarly discourse on climate justice in China. First, scholarly writing (including journal articles, conference papers, and graduate theses) on the topic of climate justice have only emerged in recent years and remain very few—72 in total since 2008. Second, when examining climate justice, more pay attention to the international aspect of the concept and highlight equity issues across the divide of developing and developed countries, while only 15 percent of the publications extend the discussion to domestic scenarios. In addition, in terms of the specific content of climate justice, about half of the papers immediately relate the concept with the issue of fair share of development, the right of development, and similar arguments. The second most popular key term associated with climate justice is equity—that is equity among nations, not equity in the domestic political context. All of these findings echo the mainstream rhetoric used in official documents and speeches.

On a slightly brighter side, new ideas are getting mentioned in some of the most recent scholarly writings on climate justice, such as applying John Rawls' theory on the "least privileged" to justify the treatment of the most climate-vulnerable countries, and linking climate issues with ethics and human rights (He 2015). A small number of publications introduced the new ideas that are galvanizing support at the global level such as the preventive approach and the "no regret" principle. Someone worth mentioning is Professor Wang Canfa, also a prominent environmentalist and the founder of the first legal assistance center for pollution victims in China, and his two journal articles devoted to climate justice (Wang and Chen 2013, 2014). Not only did he and his co-author outline the relevant approaches about climate change, they also highlighted five main principles: priority for the most vulnerable, the responsibility of the originator, equity of emissions rights, maintenance of traditional usage, and preventive measures and the "no regret" principle. Moreover, Wang ardently argued for a unified stance on climate justice by the Chinese state in both domestic and international settings, and for the principle that whatever China bargains for at the global level shall be applicable to the domestic context as well. Although some may remain critical, Wang's views are rarely found in mainstream Chinese publications and media reports. He is among the very few that are pushing for further modifications of China's policies on climate change toward the direction of going beyond mitigation and industrial pollution and including more social justice discussions.

Climate justice: still far away from our backyards?

The rise of public environmental awareness and environmental social mobilization in China is often evidenced by two separate tracks of events. One centers on environmental NGOs and prominent activists, and the other features mass protests against infamous industrial and development projects such as the anti-PX (paraxylene) protests[10] in Xiamen in the summer of 2007. Although both tracks have continued to evolve and extend their geographic scope, there is little sign of linkages between them. When such linking-up does happen, as in the case of an anti-incinerator siting movement in Panyu (2009–2012), the impact is obvious and important: not only can community leaders acquire more sophisticated tools for mobilization and policy advocacy, but also the outcome of protests can have a much more lasting effect. Governments will then be more willing to adopt recommendations from community-based NGOs and make substantial policy adjustments (Steinhardt and Wu 2016).

For a more inclusive and progressive discourse on climate change to emerge and have a real impact in China, such a merging of public intellectuals, prominent environmental activists, media, NGOs, affected communities, and concerned citizens is necessary. As explained above, there is still a substantial gap between where China stands regarding climate justice and what is becoming the consensus at the international level such as the *Bali Principles of Climate Justice*. Without organized and coherent social activism, climate politics in China will continue to be led by state agencies and focused on energy, emission reduction, and equity between developing and developed countries. There is a great need for Chinese environmental NGOs to locate and connect with climate grievances, and to articulate specific policy needs on behalf of existing and potential climate change-related victims.

Among others, one main contribution to the underdevelopment of social activism addressing climate justice is the lack of articulating the linkage between social grievances and climate change. Therefore, this chapter, based on both published research and field work experiences, intends to make a small contribution by identifying three possible socio-geographic areas where unequal resilience against natural disasters and extreme weather associated with climate change have emerged, or may emerge, and can cause significant social dismay in the medium run.

First, offshore islands and coastal regions that face potential effects of rising sea levels (Williams et al. 2016). In recent state climate documents, attention has been paid to this, and "model islands" have been identified to build up preparedness. Second, the agricultural sector and agrarian communities in general that may experience increasing and/or irregular frequencies of flooding, droughts, and other types of natural disasters (Wei, Wang, Wang, and Tatano 2015). A recent study based on data from 65 cities has tentatively concluded that central and northern regions are more vulnerable to drought than southern regions in China (Yuan et al. 2015). Third, ethnic minority communities and poor rural areas in western and southwestern provinces that need to develop sufficient capacity against extreme weather events and natural disasters (Rogers and Xue 2015; Vu and Noy 2015; Wang, Brown, and Agrawal 2013). China's western provinces are both lagging behind in economic development and rich in cultural and biological diversity. Minority communities, who usually lack access to basic infrastructures and disaster prevention mechanisms, will face unprecedented challenges associated with climate change in the decades ahead and thus need more policy support than what has been included in current climate documents.

Conclusion

Climate justice is not just an unfinished cause in China. In some sense, it has not even started so far. On the one hand, China is investing huge financial, human, and political capital into renewables, clean energy, and carbon emission reduction and moving these issues to the top of state-level agendas. Since the 1990s, the Chinese central government has committed to many international climate-related agreements and declarations, and put in place coherent sets of regulations and measurements to reform the country's industrial structures and enhance energy efficiency, pollution control, and clean energy development (Hilton and Kerr 2016; Li et al. 2016).

However, at the same time, climate politics in China clearly neglect aspects of social justice, equity in the domestic context, and civic/political rights. Moreover, while the public is much better informed about the devastating health effects of the heavy smog that has spread across all major urban centers in China in recent years, the same cannot be argued about the concrete effects on social lives of climate change. NIMBYism[11] and protests against particular polluting projects are now common in China's urban centers and wealthy neighborhoods, but few have produced lasting and widely applicable policy outcomes, or succeeded in remote and minority regions. The public is still constrained by both structural and agency factors to mobilize for broader policy topics.

Nevertheless, environmentalists and leading environmental NGOs have started to introduce the concepts of climate justice and open conversations about the linkage between climate change and socio-political reforms in general, even though such efforts need to be boosted substantially to make a real difference. To change the fact that the state plays the predominant role in deciding the nation's climate agenda and little attention is paid to climate justice in mainstream thinking about climate change in China, a lot more needs to be done about public awareness and direct public participation. Similarly, the merging of community-based protest by pollution and potential climate change victims and professional environmental NGOs needs to take place. The fate of China's people in the face of the climate crisis hangs in the balance of power among these forces moving forward.

Notes

1. The Chinese version of the Paper is available at www.ccchina.gov.cn/WebSite/CCChina/UpFile/File189.pdf. Translated by the author.

2. Chinese version of the Report available at http://files.ncsc.org.cn/www/201511/20151120095259910.pdf. Translated by the author.

3. This point does not argue whether China's climate commitments are sufficient or not.

4. http://news.ifeng.com/a/20161119/50283661_0.shtml.

5. In September 2013, during his state visit to Kazakhstan Chinese President Xi Jinping announced the *Silk Road Economic Belt* plan to strengthen economic cooperation with Central Asian and European countries. A month later, Xi proposed the *21st Century Maritime Silk Road* to enhance China's economic relations with Southeast and Western Asian countries in his address to the Indonesian parliament. Later, the Chinese government combined these two regional proposals of international development

cooperation and named it the *Belt and Road Initiative*, also known as One Belt, One Road initiative.

6. The official statistics released by the governmental agencies and All-China Environment Federation is much more than 1,996, including semi-governmental research institutions, organizations, press, and others. The China Environmental Organization Map (www.hyi.org.cn/go/) is compiled and published online by the He Yi Institute, a Beijing-based grassroots NGO led by veteran environmentalists.

7. Detailed information of both NGOs' work is available at www.cango. org/upload/files/6_%20CCAN%20

%E8%AF%84%E4%BC%B0%E6% 8A%A5%E5%91%8A.pdf and www.cycan. org respectively.

8. Available at www.ghub. org/?p=1807.

9. Email exchanges with GH research director in November 2016.

10. Masses of people came out to protest against a proposed paraxylene plan in Xiamen, Fujian province, in 2007. This was one of the first citizen outbursts in China about threats to the environment.

11. For a timeline of NIMBY protests in China, see: www.thechinastory.org/yearbooks/yearbook-2013/forum-land-law-and-protest/nimby-protests/.

24 | ENTRENCHED VULNERABILITIES: EVALUATING CLIMATE JUSTICE ACROSS DEVELOPMENT AND ADAPTATION RESPONSES IN SOUTHERN INDIA

Sumetee Pahwa Gajjar, Garima Jain, Kavya Michael, and Chandni Singh

Introduction

This chapter interrogates the notion of justice (or the lack of it) in India's domestic policies and political priorities for climate change in the context of development interventions or adaptation strategies in regions with highly vulnerable communities. Drawing on three sites in Southern India, we distil how vulnerability is created, exacerbated, or re-created, as people move: either by migrating within or across states, or by relocating within city regions. Vulnerability to climate change is socially differentiated, determined by a range of economic, political, and environmental factors, and often experienced at a local scale. However, responses to climate change are planned at national or regional levels, with rare, if any, attention paid to local spheres of governance. State interventions to address vulnerability are often crafted through "who is identified as vulnerable" and "who identifies the vulnerable," and may or may not correspond to climate policies for the country or the region. Each case illustrates that understanding vulnerability and how it is constructed is a key part of achieving climate justice. Through our analysis, we argue for *understanding vulnerability beyond the present* and translating this understanding into multi-scale, climate policy design.

Climate policy in India

India's international climate policy has always been guided by the principle of "common but differentiated responsibility"—that rich, industrialized nations are responsible for climate change through their development trajectories and consumption patterns and must, therefore, take greater responsibility for containing and responding to climate change. With this stance, India seeks climate justice at an international scale. However, the justice frame is absent from domestic policy and practice. Michael and Sreeraj (2015) empirically establish that consumption characterizes elites in the country and they benefit from high emissions while the least polluters, who are often poor and belong to marginalized socio-economic groups, live highly vulnerable lives. This is especially true in Indian cities where large impoverished populations, exposed to multiple environmental risks, are notably absent from climate policy and planning.

India began acknowledging climate change as a significant challenge for the country's development from 2007 onwards, through the national Five Year Plans, the National Action Plan on Climate Change (NAPCC) of 2008, State Action Plans on Climate Change (SAPCC) starting 2010 onwards, and most recently, the Intended Nationally Determined Contributions (INDCs) in 2015. The national climate planning process is embedded within a co-benefits framework that aims to reconcile development priorities with climate concerns. The initial implementation period for the eight sectoral NAPCC missions ran until 2017. Although envisaged to create a directional shift in the country's development trajectory, an evaluation of the NAPCC carried out in 2012 suggests that this objective remained unrealized due to a lack of long-term vision and an integrated approach (Byravan and Chella Rajan 2012).

In 2010, state governments were tasked to integrate national climate priorities with each state's development initiatives but these are yet to identify vulnerable sectors or target specific regions for adaptation (Dubash and Joseph 2015). The state planning mechanism is not designed to address inter-state socio-economic and ecological disparities. Ecological impacts of land-use change for resource extraction (mining, deforestation) and energy production often exacerbate the pre-existing vulnerabilities of populations in one region, while the consumption of materials and energy for economic growth (and related carbon emissions) may occur elsewhere.

The NITI Aayog (National Institution for Transforming India), which was appointed as successor to the Planning Commission in 2014, identified the Ministry of Environment, Forests and Climate Change (MoEFCC) as the nodal ministry for implementing Sustainable Development Goal 13 (dedicated to climate change). It also identified the NAPCC missions as the mechanisms for implementing climate interventions to help achieve specific targets under the goal. This in effect translates to a climate policy framework at the national level, leaving little room to address justice concerns across and within states and within city regions.

Cities as contributors to climate change, and as sites with high exposure to climate risks, are not taken into consideration sufficiently in the climate policies of India (Revi 2008). Successive droughts, water scarcity, and natural resource degradation have contributed to an agrarian crisis and large-scale migration from agriculture and allied sectors into informal jobs in Indian cities. Once in the city, these migrants find accommodation in informal settlements in and around the city that are poorly serviced and often illegal. Thus, Indian cities have become "highly unequal spaces economically, spatially, socially and culturally over the last two decades" (Vakulabharanam and Motiram 2012: 44).

The Jawaharlal Nehru National Urban Renewal Mission (JNNURM) of the early 2000s brought the national focus on urban development, emphasizing improvements in basic services including housing for the urban poor. This was followed by the Atal Mission for Rejuvenation and Urban Transformation (AMRUT) in 2015 with an aim to promote integrated development planning, public transportation, sustainable water management, and green spaces. However, there has been a lack of capacity among elected representatives and officials to read environmental impacts and resource conflicts as issues of climate justice, in particular the sharing of river waters across and within states, dumping of urban waste, conversion of farmland in peripheral urban areas to real estate or special economic zones. Our cases demonstrate that vulnerable groups, and the socio-economic and environmental drivers of their vulnerability, often fall outside the gambit of domestic policies aimed at addressing climate change or pushing urban transformation. Unless the geographically distributed sources of vulnerability are factored into policy formulation, climate justice will remain beyond our reach.

Understanding vulnerability through a justice lens

The impacts of environmental change fluctuate rapidly among different social groups, falling disproportionately on the marginalized, and this is one aspect of the unpredictable relationships between climate change and its outcomes (Ribot 2010). Given the unjust impact of climate change on the poor, in terms of both mitigation and adaptation (Shi et al. 2016; Klinsky, Dowlatabadi, and Mcdaniels 2012; Ziervogel et al. 2017), justice becomes a central concern for evaluating and supporting climate change responses. Marino and Ribot (2012: 323) propound that "climate change is redistribution ... As redistribution, climate change is also a matter of justice—it is about who gains and who loses as change occurs and as interventions to moderate change unfold."

The literature before 2009 on vulnerability and climate variability depict vulnerability as a linear process perceived in terms of the differential influence of climate change on people. Ribot (2014) reiterates that "understanding the causality of vulnerability has been grossly ignored." Understanding vulnerability through a structural lens raises critical questions of justice. However, climate justice as an approach to integrate concerns of social justice with challenges of climate change is a relatively new development. And as Schlosberg (2012: 449) argues, climate justice as an approach has excessively relied upon "frameworks of prevention or mitigation, or on the distribution of the costs of adapting to climate change." He also notes that the concept of mal-recognition promotes distributive injustices on the lines of class, race, income, gender, etc. and is largely absent in climate justice literature. He argues that the primary step towards recognition is to understand the status of the vulnerable and the processes through which that vulnerability is produced. This means bringing attention to a range of experiences of the vulnerable and the way that their status is, in part, socially constructed. Nussbaum (2009) notes that recognition when considered as a constituent of justice should be followed by measures that encourage participation, which should not just be limited to providing distributional access to resources, but also focus on their role in enabling the most vulnerable to lead safe lives.

At the city scale, Bulkeley (Bulkeley et al. 2014) argues that vulnerabilities associated with climate change are produced through various forms of uneven development and inequalities existing within a city. Parnell and Pieterse (2010) support this in the context of the increasing urbanization of poverty and the need to address development and universal rights in cities across scales from the micro-environmental down to the individual and household scale. This need has been recognized particularly at the city scale. As cities become "greener," develop sustainability or resilience plans, or plan for climate impacts (e.g., Rockefeller Foundation's 2017, *100 Resilient Cities* project), they do not necessarily become more equitable or inclusive. Justice at the urban scale requires attention to issues of recognition as well as rights and responsibilities. It requires recognizing existing forms of inequality and interrogating whether climate change interventions exacerbate or redress underlying structural drivers of vulnerability.

Movement of people within India—three cases

We now consider three cases of how vulnerability is created, exacerbated, or re-created, as people migrate within or across states, or are relocated within city regions. Through our analysis, we argue for *understanding vulnerability beyond the present* by delving into characteristics of everyday life and decision-making, which exclude vulnerable groups from influencing the very factors that entrench their vulnerability in places of origin, and another set of structural issues which perpetuate their vulnerability in places of destination. For example, the Bengali migrants in Case 1 are unable to change their circumstances either in their original locations or final destinations. Drawing on Schlosberg (2012), each case illustrates that recognition, participation, and agency are useful parameters towards understanding vulnerability and the ways in which it is constructed and perpetuated. Unless these parameters are addressed through policy, planning, and other forms of development intervention, climate justice will remain an elusive goal for climate responses.

Case 1: Understanding vulnerability through a structural lens: inter-state migrant waste pickers in Bangalore

This case draws on a larger study examining differential vulnerability of informal settlement dwellers in Bangalore. Out of the 32 surveyed settlements, we focus on two of the selected informal settlements, Hebbal and Marathahalli, which are located on landfill sites. A mixed method approach using quantitative and qualitative tools was used to capture in-depth household information and community dynamics. Most settlement dwellers are inter-state migrants, mainly from the state of West Bengal, working as waste pickers in Bangalore city (Michael, Deshpande, and Ziervogel 2017). Vulnerability is examined at multiple scales to capture the intricate interlinkages between "local, national, regional and global political-economic relations" (Michael, Deshpande, and Ziervogel 2017: 14).

Most of the residents in these two settlements migrated from the districts of Nadia or Murshidabad because of a combination of climatic and non-climatic factors. Both Nadia and Murshidabad are prone to extreme climatic events such as floods, heavy rainfall, and hailstorms. In addition, most of these migrants are landless agricultural laborers moving to cities as a strategy to cope with the looming agrarian crisis across rural India.

As Michael et al. 2017 argue, the vulnerabilities for such communities are replicated at both source and destination. The process of urbanization in cities like Bangalore is highly discriminatory and a large majority of distressed rural migrants find refuge only in the burgeoning informal settlements of the city. Migrant workers, usually through the help of already existing social networks, start working as waste pickers under a community leader called "thekedar" ("contractor") and lead fragile lives in settlements.

Hebbal and Marathahalli settlements are located on landfill sites close to open drains. The dwelling spaces of the migrants also double up as their work stations where they dump and sort all the waste collected. Being inter-state migrants, they are deprived of support provided by the state and do not possess any identity cards such as ration cards or voter IDs from the state of Karnataka. These communities are also under constant threat of eviction and lack access to basic services such as water, sanitation, and electricity. While local waste pickers possess identity cards that legally permit them to collect waste from various unauthorized dumping grounds in the city, migrant communities are not entitled to such cards. Hence, migrants exercising their right for a livelihood in Bangalore are considered illegal by governing authorities. Temperature rises, changing precipitation patterns, and proximity to drains have also led to dengue outbreaks, which translate into livelihood losses. In addition, they face significant income losses during floods because the water soaks the waste they collect. Although these waste pickers perform an important ecosystem service for the city, they are virtually unrecognized by the city and the stigma associated with their occupation has a detrimental impact on their sense of dignity.

Case 2: Tracing historic marginalization that shapes present-day vulnerability: intra-state migration in Karnataka

Migration within the state of Karnataka is also characterized by climate variability (mainly drought) and filtered through social stratifications. Rain-fed agriculture is the primary source of livelihood in rural Karnataka and increasing climatic variability, fluctuating market prices, eroding natural resource bases, and shrinking landholdings are driving people to migrate, either seasonally or permanently, to urban centers (Basu and Bazaz 2015; Singh, Basu, and Srinivas 2016). Our research in two source districts (Kolar in the southeast and Kalaburgi in the north) and one receiving district (Bangalore) shows that movement for livelihoods encompasses a range of strategies: (1) seasonal migration to metropolitan cities such as Hyderabad, Mumbai, Pune, and Bangalore in the lean months for supplementing agrarian incomes, (2) permanent migration, typically to Bangalore, (3) commuting to smaller towns (usually the district headquarters) on a daily basis, and (4) opportunity-based movements which are not necessarily temporary (such as moving to the city for education or changing locations for marriage).

To understand migration patterns, drivers and outcomes data were collected through a household survey (n=820), 26 gender-differentiated focus group discussions, and 17 in-depth life histories of migrant and non-migrant families at source and destination. The analysis used a frame of intersectionality and temporality to assess whether past socio-political marginalization and trajectories of exclusion mediate experiences of vulnerability.

Historical patterns of social exclusion manifest in present-day vulnerability. For example, tribal farmers have smaller asset bases (marginal landholdings, low water availability, poor soil to till in), lesser agency (low participation in local development initiatives), and poorer financial and administrative connectivity (weak connections with formal banking systems, low proximity to block or district headquarters). Such pre-existing inequities follow households to shape their responses to climatic and non-climatic risks. For example, we found that tribal households tend to move out of agriculture more often because of smaller landholdings and often move into unsafe and temporary livelihoods in cities where issues of access to public services, identity, and safety exacerbate their vulnerability. Seen through an intersectionality lens, gendered norms mediate other variables such as age, caste, and education levels to shape the capacities and opportunities of individuals. Thus, women migrants often reproduce rural norms of what work a woman should do and such norms shift, based on length of time in city, education levels, and household needs. Women from upper castes tended to stay in villages, performing roles that maintain exclusivity, as distinct from women from Scheduled Castes who travel to work in nearby garment factories and sell produce in local markets. Thus, the case of migration at an intra-state level demonstrates how vulnerability is socially differentiated and embedded in historically shaped structural inequities, for example, of gender.

Case 3: Assessing creation of new climate risks at multiple scales— post-disaster resettlement of lower income groups in flood-prone areas of Chennai

In the last two decades, Chennai the capital of the state of Tamil Nadu, has faced several climatic and tectonic hazards including floods, cyclones, drought, and a tsunami. Following these disasters, Tamil Nadu has often assumed the role of a welfare state and compensated the affected lower income group communities living in hazardous areas. Such compensation has taken various forms including housing allocation for the affected communities, although the housing is located several kilometers away from their current sources of livelihood.

After the Indian Ocean tsunami in 2004, the affected fishing communities near Santhome Church, for instance, were allocated housing in Kannagi Nagar and Semmencherry. After the 2015 floods, more families from the river flood plains of Saidapet have also been moved to Kannagi Nagar (now expanded over the years) and Perumbakkam close to Semmencherry. All these locations are within the low-lying area of the Pallikarnai Marsh which is flooded every monsoon. These locations are far from the city and not well connected to essential socio-economic services, affecting people's everyday life and livelihoods (Jain, Singh, Coelho, and Malladi 2017). Relocation to under-serviced areas is also creating greater environmental risks for the city. To understand vulnerability to climate change impacts at household level and at city scale, and to critically examine state-led actions for vulnerability and risk reduction, we conducted in-depth interviews with affected and relocated households (n=60) and around 30 expert interviews and consultations with planners, academics, and social and environmental activists.

Pallikarnai Marsh is a critical wetland for the city and the entire state of Tamil Nadu. Over the last two decades, large segments of the wetland have been converted into residential and industrial zones (Vencatesan, Daniels, Jayaseelan, and Karthick 2014). A significant portion of this development is governed and delivered by the state for various purposes, including affordable housing for poor, inner-city slum dwellers. Other state-led developments in the marsh include a waste dump-yard. Many experts claim that encroachment of this marsh is one of the primary reasons for flooding in the city as well as the slowing down of groundwater recharge, leading to water scarcity and increased salinity of groundwater (Vencatesan, Daniels, Jayaseelan, and Karthick 2014). While relocation of human settlements is justified to reduce hazards, other forms of development such as an international airport and large industrial complexes are being undertaken on "hazardous" locations. In the past, high-to middle-income group housing and commercial developments have also been carried out on reclaimed lake beds.

Critically examining the construction and drivers of vulnerability

The case of inter-state migrant waste pickers in Bangalore city critically points to the need for bringing an urban climate justice lens to climate adaptation. The vulnerability experienced by migrant waste pickers is highly structural in nature and is a combination of socio-economic

Table 24.1 Typologies of Movements and Their Interaction with Key Constituents of Climate Justice

Movement typologies Key constituents of climate justice	Intra-city state-led relocation in Chennai (planned)	Intra-state migration in Karnataka (autonomous)	Inter-state migration to Bangalore (autonomous)
Underlying environmental and structural drivers of vulnerability at source	Riverine and coastal flood risk and tectonic hazards intersecting with pre-existing socio-economic deprivation, particularly lack of tenure security.	Water scarcity and drought, natural resource degradation. Social differentiation based on caste and asset bases (especially land ownership, access to irrigation).	Flooding and cyclone incidence. Agrarian crisis leading to unviable livelihood options, socio-economic marginalization.
Who is vulnerable?	Socio-economically and politically weaker sections with limited habitation choices within the city, and often living in under-developed locations, exposed to climatic and non-climatic hazards.	Landless households; households belonging to Scheduled Castes and tribes (recognized as marginalized as per the Indian constitution), and households with lower-value assets because of historical marginalization.	Landless agricultural laborers from West Bengal, mostly Muslims, who have migrated to Bangalore and are working as waste pickers.
Trigger for movement	2004 Indian Ocean tsunami and 2015 December floods triggered state-led relocation projects.	1972, 2010, 2015 droughts, falling groundwater levels since 2005, and growing discontent with agriculture as a livelihood.	2002 cyclone triggered mass migration to Bangalore.

Recognition	"Identifying" disaster-affected communities living in informal settlements as needy of relocation is a form of misrecognition (Fraser 2003; Schlosberg 2012).	Caste-based differences in assets and capabilities, while recognized by the government, do not address structural drivers of vulnerability.	Lack of tenureship rights and access to basic services.
	Unabated infrastructure and real estate development, which continues in environmentally sensitive/hazard-prone locations.	Social norms around caste and class used as tools of "social subordination" (Fraser 2003; Schlosberg 2012).	Invisible to the municipal authorities due to the illegality of their settlements. Absence of identity cards makes their livelihood illegal, leading to misrecognition (Schlosberg 2012) and injured identities (Michael et al. 2017).
Participation	No participation in the decision-making or implementation of the resettlement process. After moving, people report losing prior socio-political capital to participate in meeting their immediate needs.	Participation in the destination is limited but language and length of stay in city improves ability to participate to some extent. Migrants visit their villages regularly to participate in local politics, for subsidized food rations, medical support etc.	Lack of recognition and their illegal status limits their negotiating power to participate in local politics or get access to basic services like water, electricity, etc. in their settlements.
Agency around moving decision	Despite existing resistance, people are moved post disasters under the pretext of safeguarding them against future environmental hazards, leaving no alternatives.	Seemingly autonomous decisions to move are shrouding the role of the state in creating conditions that necessitate movement (e.g., through unsupportive agricultural policies).	Decision to move is shaped by the larger political economy of agrarian distress and socio-economic marginalization in West Bengal. Agency is limited to choice over where to migrate.

(continued)

Table 24.1 *(continued)*

Key constituents of climate justice / Movement typologies	Intra-city state-led relocation in Chennai (planned)	Intra-state migration in Karnataka (autonomous)	Inter-state migration to Bangalore (autonomous)
Circumstances around moving	State-led initiatives of building housing stock when funds are available, and using disasters as moments of opportunity to relocate communities from informal settlements within inner-city locations, to peripheral and environmentally sensitive locations.	Social networks, proximity to road facilitate migration. Mostly men tend to move to cities (in construction) while women move to neighboring rural areas (for agricultural labor).	Existing social networks in the city help to facilitate movements. The particular livelihood migrants are pursuing in Bangalore wouldn't have been possible in their native lands because of the social stigma associated with it.
Vulnerability at destination	Despite state-led relocation, physical risks to floods continue, and new economic and socio-political risks emerge, such as lack of access to viable livelihood options, social infrastructure, and accessing governance systems. Lack of tenureship rights persists.

Environmental vulnerability is also created in the process for the city by building on wetlands. | At destination, people experience difficulties in accessing water and services such as transportation and healthcare, dealing with local flooding, especially if living in blue tent settlements, and getting jobs.

Although income in the city is higher, people's social networks and participation in the city is very low. | Hazardous and insecure livelihood. Lack of access to basic services. Climatic factors like flooding and erratic rainfall patterns disrupting livelihoods and living conditions. |
| **What are the injustices?** | Vulnerabilities to climate risk are re-constructed, as social subordination is institutionalized by disallowing the developmental rights of the vulnerable. The rights of the relocated communities to the city, and to determine their own future are denied. | Historically determined injustices based on cultural and political exclusions drive economic inequality and shape current capabilities to diversify livelihoods. | Structural vulnerabilities like socio-economic marginalization deny opportunities to sustain in their native places.

In the city even though they perform an essential ecosystem service they are denied the right to live in a clean and safe environment. |

Source: Authors' analysis based on Schlosberg (2012) and Fraser (2003).

and ecological marginalization experienced in their homelands that gets replicated in the city. The insecure and hazardous livelihood in the city also leads to injured identities due to rejection by society and raises critical questions of recognition. Climate change acts as a crisis catalyst and accentuates these vulnerabilities by altering livelihoods and living conditions on a daily basis. Even though they perform an important ecosystem service by collecting and processing a large amount of the city's garbage, the services of waste pickers often remain invisible in the eyes of the city, thus raising a major question of justice.

In the case of intra-state migrants in Karnataka, historical patterns of unequal development are perpetuated *within* and *between* communities. However, acknowledging past structural conditions as drivers of current injustices requires us to move beyond justice merely in terms of distribution of goods, towards justice that helps build people's capabilities to function in what they value (Bulkeley et al. 2014; Schlosberg 2012). Seen from a "justice to meet capabilities approach" (Schlosberg 2012), intra-state migrants in Karnataka highlight the necessity of recognition, of recognizing structural patterns of advantage and disadvantage, and subsequent disparities in people's ability to respond to multiple risks.

The Chennai relocation case reinforces issues of recognition as well—who sets the definition of an "encroacher," who gets identified as one and who does not. This illustrates Fraser's (Fraser and Honneth 2003) "institutionalized social subordination"—where a certain set of people are continually marginalized by the system. In this case, it is the poor who are not able to formally own land or housing in the city. While the policy to move people may be well-intentioned, the processes of governance and decision-making at the time and site of relocation, and service levels therein, are indeed questionable. When people are asked to move at the time of a disaster, their ability to participate effectively in decision-making towards their future well-being is considerably compromised. Following the capabilities framework (Bulkeley et al. 2014; Schlosberg 2012), developmental and environmental rights are harmed when moving people from a location of physical risk. Relocated people lose the little socio-economic and political capital they may have built and their access to the natural resources critical for their livelihoods such as fishing. At the same time, displaced communities stand to gain nothing from the conversion of the land they had hitherto inhabited, a conversion that is often for building city infrastructure or for commercial or real estate purposes. This is because they can no longer access the city, having been moved to a peripheral site.

Conclusion

Each case illustrates the movement of people because of social, economic, and environmental factors exacerbated by climate change. The cases show how vulnerability is recreated for displaced communities despite relocation, be it through loss of socio-economic capital and political agency during or after resettlement (Chennai resettlement) or through a lack of recognition in a people's role in a city's economic and environmental functions (Bangalore waste pickers and Karnataka migrants predominantly engaged in the informal economy).

As our cases exemplify, the normative goal of climate justice (espoused in the stance for international negotiations and national policies) can be extremely challenging to maintain when it comes to ensuring climate justice for people as part of households and communities in specific local contexts. Therefore, although policies for adaptation recommend maintenance of ecosystem services through conservation of urban nature, the environment is readily compromised by urbanization and the relocation of people on ecologically sensitive land (Pallikarnai marshland). Similarly, although the AMRUT mission aims to improve infrastructure and provide basic services to underprivileged urban residents, the legal status of migrants prevents their access to well-intentioned programs (Bangalore). Where people are identified for upliftment, the programs may be ill-designed and exacerbate vulnerability (relocation to peri-urban Chennai). Clearly, if progress towards climate justice is to be made, structural and environmental drivers of vulnerability need to be understood.

25 | SELF-INTEREST TRUMPS GLOBAL CLIMATE ACTION IN AFRICA AND ELSEWHERE

Nnimmo Bassey

The 1992 Rio Summit ought to have been a great opportunity to lay out the road map for solving the mounting environmental changes of our time. It was not. The possibility was thwarted by the insistence by George H. Bush, the president of the US, that the American way of life was not negotiable (Deen 2012). The moment national interest trumped the global necessity, multilateral efforts to tackle critical environmental challenges got booby trapped.

Roll forward to 1997 at Kyoto where the Conference of Parties (COP) to the United Nations Framework Convention on Climate Change (UNFCCC) met. The resulting *Kyoto Protocol* (KP) was touted as a breakthrough agreement, being the first legally binding instrument for tackling global warming. As deficient as the KP stood, COP15 held in Copenhagen could not reach an agreement on a second commitment period of the *Protocol*. We should add that even if a second commitment period had been reached at Copenhagen, the aggregate effect of its implementation would not have seen the world off the path to climate catastrophe.

The outcome of 1997 had minimal requirements for emissions reduction required by science. It also brought with it rather upsetting templates for carbon offsetting and accompanying market mechanisms. The redeeming aspect of Kyoto that could not be killed was its justice undergird: the common but differentiated responsibilities according to capacities.

What trumped Kyoto? Selfish national interest in the Round of 1992

If Kyoto was cobbled with minimal binding requirements, the Copenhagen COP shamelessly concretized the setting aside of actions along the lines of global interest in favor of non-binding commitments.

It was at that COP that arbitrary targets were set for the mobilization of funds for a so-called Green Climate Fund (GCF)—a palliative to soothe the nerves of vulnerable poor nations, a carrot held far in front of a horse to keep it galloping on even when dead tired. The wealthier nations were to mobilize US$10 billion from 2010 to 2012 and raise this to US$100 billion per year from 2020.

One hundred billion US dollars seems a huge outlay that is hard to raise, but think about the hundreds of billions of dollars spent annually to maintain nuclear weapons. When we also think about the trillions of dollars spent annually on warfare by the rich nations, the seriousness to tackle global warming becomes suspect. The means of mobilizing the funds remain complicated and contested with many insisting that it should all be new funds rather than being tied to foreign development aid, private finance, and the like.

Fast forward to the Paris COP21 and the celebrated *Agreement* (UNFCCC 2015) that emerged from it. The *Agreement* was applauded and rapidly endorsed by virtually all nations—both the polluters and the polluted. It was hailed as the first time the entire world recognized the need for global climate action. But, be that as it were, the *Paris Agreement* merely elevated to global policy the doctrine of national ways of life not in any way being challenged by a global imperative. The major plank of the *Paris Agreement* is the (Intended) Nationally Determined Contributions or NDCs. With the NDCs, nations crafted and submitted what they felt they would do to reduce emissions and curtail harms to the climate. The NDCs in reality lock away the requirements of science that collective actions should not be taken at the whims of nations but as computed necessities that would keep global temperatures from rising inexorably. In fact, for some of the nations, the implementation of aspects of their NDCs are dependent on the availability of finance and access to needed technology. This is an obvious stumbling block.

The *Paris Agreement* was, in a sense, concocted as a sedative. It stated the objective of keeping global temperature rise to 1.5 degrees or "well below" 2 degrees Celsius above preindustrial levels. The analysis of the Intergovernmental Panel on Climate Change (IPCC) quickly showed that if all nations faithfully implemented their NDCs global temperatures would rise by up to three degrees by the end of this century. Sadly, that grim reality never deterred the applause for having a global agreement on climate change. By 2012 a study by the Congress of South African Trade Unions (COSATU) had already underscored the enormous economic impacts rising temperatures would have on Africa. The study found that by 2040, a 1.5°C temperature rise would cost Africa 1.7 percent of its Gross Domestic Product (GDP) while a 2°C scenario would cost the continent 3.4 percent of its GDP by 2060. An alarming 10 percent would be eaten up by 2100 if the temperature rose by 4.1°C (Nel 2015). A COSATU official is also on record to have said:

> Climate change is not a middle class preoccupation. It will affect unemployment, economic structures and industrialization. Already there are over 150 million "climate refugees" in the world displaced by drought, failed crops, floods and rising sea levels. In addition, 262 million people are affected yearly by climate disasters. It is the working class, the poor and developing countries that are most affected. (Nel 2015)

He went on to add:

> We link environmental challenges with socioeconomic challenges. Cosatu shows workers that there is a direct correlation between the minerals-energy complex and increased greenhouse gas emissions. We talk to workers on the shop floor at the big companies so they can participate in debates and decisions. (Nel 2015)

We should not forget the memorable scene at COP15 where the African negotiator could not imagine an agreement hinged on a 2°C temperature change. At that COP, Lumumba Di-Aping, the lead negotiator for the Group of 77 and China denounced the 2°C warming target as "certain death for Africa." He also called it a type of "climate fascism" that was being forced on Africa. He wondered why Africa was asked to sign an agreement that would permit warming in exchange for US$10 billion, and also being asked to celebrate such a deal (Bond 2012). What has changed since then?

It was clear already that the world had lost the collective resolve to tackle global environmental changes frontally at the Earth Summit of 1992. We have simply hobbled along from that time. Rio+20 simply raised the bar by enthroning the doctrine of the "Green Economy," insisting that ecological issues can best be tackled via the market and that unless monetary economic value is placed on Nature any thoughts of protection or preservation would be merely wishful (Clemençon 2012).

The justice base is being blurred by the toss of coins in the way of an opaque GCF. In exchange for silver coins in place of colonial beads and mirrors, impacted countries are forced to absorb whatever equivalent of pollution emitted by industrial nations enables the latter to maintain their polluting production and consumption lifestyles. To ensure these false climate actions, the pursuit of foreign exchange and the much-loved foreign direct investment, our governments do not hesitate to criminalize and displace individuals and communities that are unwilling to succumb to such questionable ideas and projects.

The market has been an excellent tool for the exploitative integration of Africa into the global loop. Promoters of market environmentalism brought up the concept of carbon offsets, clean development mechanisms (CDM), and reducing emissions from deforestation and degradation (REDD). The commodification of Nature, the voluntary actions on climate change, and the rise of market environmentalism all complicate the existential challenges facing Small Island States, African nations, and others. These states have a duty not to celebrate "intended" climate actions that do not respond to science. Anything else is nothing but the singing of one's funeral dirge.

Fossil fixation

The climate future will be largely determined by how humankind's fossil fixation is handled. Will the lure of financial profit from the extraction and burning of coal, crude oil, and gas trump the realities of the fact that we are more or less at the precipice? Would the harm already done to vulnerable Small Island States, Pacific, African, Asian, Arctic communities, and others be ignored or would the world eventually wake up to face the menace? Experts have stated that to keep within the carbon budget and hope to avoid more than $2°C$ temperature rise above preindustrial levels, 80 percent of known fossil fuel reserves must be left unburned. It is incomprehensible that with this knowledge, the UNFCCC has not deemed it fit to take this into account in the negotiations. It appears that the word "fossil" is anathema in the negotiation process and it is nowhere to be found in agreements that have been reached. This situation may be orchestrated by the fact that fossil fuel companies are embedded in the negotiation processes.

The fact that oil companies have invested resources towards refuting findings of climate science linking their products to global warming indicates that with the continued presence of fossil fuel companies in the UNFCCC process, real climate action may *never* be on the cards. Rather than cutting back on dependence on fossil fuels, oil companies are investing in exploration for new reserves and in sowing the fiction that the world will continue to be dependent on fossil fuels as a dominant energy source for the foreseeable future. If that is the long haul and the foreseeable future for the fossil fuel industry, their sights must be myopic. The trend is beginning to emerge that points in the contrary direction (Akpan and Eboh 2017). For example, some countries including India, China, Japan, Norway, and the UK have announced that they would not be manufacturing or selling cars with internal combustion engines in the coming decades (Beament and Agerholm 2017). We are also seeing a global scaling back on coal-fired power plants and a rise in the uptake of renewable sources for electricity generation. The shift is clearly away from a lock-in on the fossil path.

Fossil resources have served as such cheap means of amassing profit for corporations and their shareholders because of the rampant externalization of social and environmental costs by the players in the sector. While ignoring the unconscionable impacts wrought on poor communities and on the climate, fossil fuel companies sell the narrative that technology can save the climate. Techno-fixes include carbon capture and storage, genetic engineering, and geoengineering. Another argument is that countries depending on fossil fuel resources for revenue would not readily leave such resources in the ground. They also claim that it would be foolhardy of any nation to refuse to extract fossil fuels because of climate impacts—according to them, if they do not extract, someone else will. The argument is that as long as there is a demand there will always be a supply.

There have also been calls that in the transition from fossil fuels, the rich industrialized countries should take the lead, while the nations that contributed the least to global warming should be allowed to extract and get a fair share of the remaining carbon budget (Holz, Kartha, and Athanasiou 2017). This argument sounds reasonable on the justice platform when it is considered that the industrialized countries are the ones that have taken up virtually all of the 80 percent of the carbon budget that resulted in the crisis confronting us now. That argument has not been bought by the rich nations who are not ready to compromise their lifestyles marked by conspicuous consumption and rampant wastage.

With leaders of powerful nations unwilling to act, the fossil fuel industries persist in the search for resources through extreme means such as fracking and in precarious locations such as in the deep seas and the Arctic.

There are some concerns that cannot be ignored with regard to the transition from fossil fuels. One such concern has to do with job losses.

What would happen to the workers in the dirty energy sector? Would they simply be thrown out of work? A consideration of this question led labor unions such as the International Trade Union Confederation (ITUC) to call for a *just transition*. Part of what that entails is that workers in affected sectors should be offered the opportunity to learn new skills in order to transition into new jobs as the world transitions away from dirty energy. The ITUC-Africa believes that

> the issue of climate change and its devastating effects cannot be divorced from the current production and consumption models. These are modes of production that place their need for profits higher than the needs of people especially the poor. The increase in production and consumption has a devastating effect on the ecosystem. The natural environment and resources, the air and soil is being polluted, water levels are dropping and chance to access food grown in a healthy and natural environment is shrinking at an alarming rate. (ITUC-Africa 2015)

The confederation also argues that

> while the transition to low carbon is crucial, such interventions must be done in a fair and through a just process. In this regard, it is of utmost importance that the developed countries should lead in reducing emissions because they are responsible for much of the environmental and climatic crises the world is facing at the moment. (ITUC-Africa 2015)

The transition is inevitable. The question is how soon and on what scale it will take place. Although he was writing on the need to shift from nuclear power to renewable, the wisdom of Hermann Scheer is worth noting here:

> If the transition from the nuclear and fossil to renewable energy is only carried out in a piecemeal and gradual manner, then it is highly likely that world civilization will be thrown into a staggering crisis affecting everyone and everything: dramatic climate change threatens to make entire habitats unfit to live in and to trigger mass misery and the migration of hundreds of millions of people. (Scheer 2012: 3)

Another sticky issue in climate negotiations is the need for financial support for vulnerable nations. We cannot deny the fact that huge financial outlays are needed for climate mitigation and adaptation measures in the vulnerable developing countries such as the least developed countries, small island developing states, and Africa. Extreme weather events have wreaked largescale havoc in countries such as the Philippines and Pakistan. Floods and droughts have become frequent experiences in Africa. Their intensity increases by the year. From COP15 to COP21, the UNFCCC process has thrown up a Climate Fund that does not appear to recognize the gravity of the problem. The same unmet promise has been made in Paris to raise US$10 billion per year over a three-year period and to increase that to US$100 billion a year from 2020. This funding is expected to come from a wide variety of sources including public and private, bilateral and multilateral, and other alternative sources of finance. The system has presented US$10 billion and US$100 billion as huge inconvenient amounts to be coughed up by rich and powerful industrialized nations. There are possibilities that these nations will wriggle out of responsibility by utilizing the "creative" financial accounting to include loans and grants made in other circumstances to the vulnerable nations.

It is telling that the United Nations Environment Programme (UNEP) believes that 85 percent of the finance required to make the shift will be derived from private investors. "The commodification of pollution is inspired by the rationale of market efficiency: major polluters issued with permits are incentivized to emit less, thereby enabling them to make a profit selling excess permits to those less efficient" (Sharife 2011: 157). The logic that the markets that created the climate crisis hold the key to solving the problem while carrying on with business as usual is the peculiar logic of official climate strategies.

To reiterate, the question that we need to examine is why it is so difficult to find money for climate finance whereas there do not appear to be any difficulties in finding finance to wage war—a destructive exercise that lowers the climate resilience capacities of those at the receiving end of military aggression. Since 2001 there has been a rise in military expenditure by the industrialized nations. Their war budget rose to over US$1.7 trillion in 2011. It is obvious that a fraction of that amount would save lives and help combat the ravages of global warming if applied for those purposes. If just 25 percent of the war budgets were to be set aside for climate mitigation/adaptation measures there would be US$434.5 billion in the kitty and the world would be the better for it. Just consider that one stealth bomber costs a whopping US$1 billion (Bassey 2016).

Can leadership from below solve the logjam?

After more than two decades of meetings and negotiations, the UNFCCC has not been able to reach an agreement that demands real climate action and a recognition of the fact that there is a high level of urgency in the climate crisis. The much-celebrated *Paris Agreement* only requires nations to voluntarily suggest what measures they would take to curb the release of greenhouse gases into the atmosphere. The voluntary emissions reduction platform is a huge climate casino where everything is left to chance. Little wonder that the *Paris Agreement* speaks of a target of well below 2°C temperature rise while working towards a 1.5°C temperature rise above preindustrial levels, while the NDCs add up to a temperature path that far exceeds the stated targets. That is what happens when things are left to chance. Is the casino system the inevitable pathway?

The peoples of this world are not ready to gamble with the future of the planet and her children. At the epochal *Peoples' Agreement* (2010) reached at the World People's Conference on Climate Change and the Rights of Mother Earth in Cochabamba in April 2010, the peoples of the world made a clear demand that countries cut their emissions by at least 50 percent at source in the second commitment period of the *Kyoto Protocol* (2013–2020), without recourse to offsets and other carbon trading schemes. Regrettably, that commitment was never met. Considering the means of realizing the needed climate finance, the *Peoples' Agreement* demanded that developed countries should commit 6 percent of their GDP to finance adaptation and mitigation needs. The conference also stated the fact that there is a climate debt that must be recognized and paid. While not seeing the payment of climate debt as a mere demand for reparations, the conference affirmed that it is a mechanism for the decolonizing of the atmospheric space and the redistributing of the carbon. It also recognized the climate debt as a means towards obligating humans to take actions to restore disrupted natural cycles of Nature (Bassey 2012: 109–11).

The lesson of the *Peoples' Agreement* is that while political leaders work within short cycles of time determined by the electoral calendars, the peoples of this world look ahead into tomorrow and the future that coming generations will face.

Climate change is a crisis of the global commons and while nations can take actions based on their perceived national needs, the need for scientifically determined global actions must *not* be ignored. This is the crack into which global negotiations on climate change have slipped—a crack that negates the dictates of science. It questions the rationale behind having the IPCC—a body that assesses the latest science and brings up scenarios that predict what would be the results of global actions and inactions. While leaders dither, the climate crisis intensifies. With rising tides, receding coastlines, increased desertification, droughts, floods, conflicts, famines, and critical challenges to food production, it should be clear that global warming ultimately cannot be tackled on the basis of national self-interest or nationally determined actions without an eye on the aggregate outcomes. Unfortunately, with climate deniers getting into high office and seeking ways to banish the mention of the phrase "climate change," the world may slide over the precipice unless the peoples arise and force climate action on those they elected into leadership positions. Given all this—climate inaction combined with resilience reduction through warfare and the accompanying forceful displacement of populations—future realities could not be starker.

26 | FRONTLINES, INTERSECTIONS, AND CREATIVITY: THE GROWTH OF THE NORTH AMERICAN CLIMATE JUSTICE MOVEMENT

John Foran, Corrie Grosse, and Brad Hornick

On September 21, 2014, over 400,000 people marched in New York City, demanding that the governments of the world take action on climate change. In 2015, the movement forced the words "climate justice" into the woefully inadequate *Paris Agreement*, though qualified with the dismissive words "as some call it." And since then, the global movement has continued to grow—in strength, in numbers, in visibility, in diversity.

While the movement is global, with strongholds in all corners of the Earth, North America has loomed large in this story. From blocking pipelines to blockading coal and oil bomb trains, from keeping the fossil fuel industry from billions of dollars in investments by universities, churches, and unions to reinvesting in communities' own clean energy cooperatives, it has proven creative, determined, and powerful. Composed of students, Indigenous nations, women, folks of all gender and sexual identities, union members, frontline communities of color, young people, old people, and people of every age in between (and as yet unborn generations), it is wide-ranging, resourceful, and broad.

In this essay, we outline the contours of the movement in Canada and the US and consider the following questions: How does climate justice differ from the solutions that our governments offer? What are the key strengths of climate justice in facing the climate crisis? How is the climate justice movement in North America distinct and why should it lead climate action? How can it move forward on all levels?

Canada: Trudeau and climate—meet the new boss, same as the old boss

For many Canadian climate activists, the "fresh new face" of Canada at the Paris COP21 climate meetings and the role Canada played in driving the 1.5 degrees Celsius agenda at the meeting held out hope for a national course change. From 2005 to 2015 the country was headed by Prime Minister Stephen Harper, winner of such dubious distinctions as "Lifetime Un-achievement Fossil Award," whose government extinguished multiple Canadian environmental laws, muzzled climate scientists, harassed environmental NGOs, created "anti-terrorism" legislation targeting First Nations and other pipeline activists, and generally introduced regressive and reactionary social policy while promoting Canada as the world's new "petro-state."

"A lot has changed," says Naomi Klein, pointing to "a new federal government, a new international reputation, a new tone around First Nations and the environment. But when it comes to concrete action on lowering emissions and respecting land rights much remains the same" (2016a). Prime Minister Justin Trudeau's political capital derives largely from distinguishing himself from Harper's vision of a fossil fuel-driven economy that relies on the decimation of environmental regulations and colonial expansion deeper into First Nation territory. Yet the politics of the Liberal government, like that of Harper's Conservatives, is based firmly in a neo-liberal order, with the structural economic imperatives of an expansionist capitalist and colonial dynamic.

Trudeau has adopted the same carbon emissions reduction target as Harper: 30 percent below 2005 levels by 2030, the weakest goals within the wealthy nations of the G7. He intends to meet climate promises by imposing a national carbon tax on the Canadian provinces, with a starting price of C$10/ton beginning in 2018 and increasing to C$50/ton by 2022 (Campon-Smith 2016). But at the same time, he green-lighted the massive Pacific Northwest fracking / liquefied natural gas (LNG) project, as well as Woodfibre LNG in British Columbia, and as we write this, the expansion of the Kinder Morgan Trans Mountain pipeline, designed to transport tar sands oil from Alberta to port in British Columbia (Knight 2016). Viewed in this light, Trudeau's strategy is clear: introduce half-measures as a cover for the uninterrupted extraction and transportation of gas, coal, and oil. The maneuver became transparent when the province of Alberta's Premier Rachel Notley insisted her government would only support the new federal carbon pricing if a new tar sands pipeline were approved (CBC News 2016).

Carbon pricing and other market mechanisms have always been a false solution and a cover for inaction, from the beginnings of the United Nations Framework Convention on Climate Change (UNFCCC) to the present orchestration of political duplicity by the Trudeau government (Smith, R. 2014). In a recent article entitled "Why We Need a Carbon

Tax, and Why It Won't Be Enough," leading North American environmentalist Bill McKibben (2016b) makes the case that a climate fight, rather than mere market tinkering, is what is really required, akin to a "wartime mobilization" in which all resources at every level of society are mobilized.

The electoral honeymoon now over, Trudeau is perilously treading public relations waters, caught between rhetoric that appeases both sides with contradictory "short-term economic goals and longer-term environmental ones" according to an editorial in Canada's largest daily (*The Star* 2016). Building more pipelines is scientifically incompatible with Canada's climate change commitments, not to mention ecological survival itself, yet Trudeau approved Kinder Morgan's Trans Mountain pipeline, while making the absurd claim that Canada actually needs new tar sands pipelines to finance its transition to a green economy (McSheffrey 2016). The business press applauds Trudeau's actions, arguing that he will out-smart Harper, and successfully build at least two major tar sands pipelines by "building more solid relationships with First Nations" (Hislop 2016).

Calling Trudeau's bluff, and leaving no room for ambiguity, in the late summer of 2016 over 85 First Nations signed a new Treaty Alliance (2016) to stop the expansion of Alberta's tar sands, mobilizing communities along the routes of five major planned pipelines expansions. The Indigenous tar sands treaty, inspired by the historic campaign of the Standing Rock Sioux Tribe that brought together over 200 Indigenous tribes protesting the building of the Dakota Access Pipeline, is described by one news source as "Trudeau's worst nightmare" (Cox 2016). Along with the abiding radicalism of numerous individual frontline First Nations Land Defense actions, this initiative is a forceful corrective to the ambivalence of "progressive" Canadian political parties and their environmental NGO collaborators.

Equally unambiguous are the multiple, dispersed actions that directly challenge the extraction of more gas, coal, and oil, such as the arrest of Montreal activists who shut down National Energy Board meetings on the Energy East Pipeline. The action led to the suspension of the hearings, the firing of its commissioners due to conflict of interest (Shingler and Smith 2016), and the arrest of 99 young student activists (CTV Ottawa 2016). Other examples include local struggles on Burnaby Mountain against Kinder Morgan, the Toronto Line-9 protests, the Mi'kmaq anti-fracking protests, the 350.org- and Greenpeace-sponsored "kayaktivist" events, protests against the massive Site-C and Muskrat

Falls hydroelectric projects, the work of the Canadian Union of Postal Workers to transform the role of the national post office to a "green hub," and the ongoing land defense movements at Unist'ot'en (2016), Madii 'Lii (2016), Lelu Island (2016), and Sutikalh (2014).

The climate justice movement in Canada has a special role to play. It must harness the sense of urgency and need for wholesale change in the light of imminent catastrophic crisis, and push all political forces, especially those on the "Left" such as the New Democratic Party and Green Party, towards radicalization. Avi Lewis, champion of the 2015 Leap Manifesto, argues that authentic government action is required to avert the mounting climate crisis. Causing a minor ruckus at the 2016 New Democratic Convention, the Leap (2015) aims at more profound changes to the Canadian economy and society. It includes the exhortation to simply leave gas, coal, and oil in the ground, along with the promotion of various other progressive social policies, including ending international corporate trade deals, implementing a national childcare program, and a universal basic annual income.

Yet even the progressive, reformist, and gradualist Leap program, like its counterpart, the US Green New Deal, needs to be explicit about urgent systemic change: curbing the relentless imperatives of capitalist expansion, rejecting the further commodification of nature through green capitalism, and exposing the inadequacy of a new productivism based on new energy technologies. Inevitably this demands an ecosocialist agenda that calls for substantial reduction of economic throughput, and that fights inequality and the control of the "One Percent" (which may be extended to the top 10 or even 20 percent (Reeves 2017)), by putting collective interests above private property and arguing that electoral "democracy" in its formal, procedural, and liberal sense is no substitute for real political action and participation, or as Baber and Bartlett propose (see Chapter 17, this volume), deliberative environmental democracy might be an alternative.

The United States

In the United States, climate justice activists face a similar dearth of truly progressive elected leaders. Under the Obama presidency, the United States became the number one global producer of fossil fuels (US Energy Information Administration 2016), a dubious distinction compounded by the export of disastrous fracking technology and the accompanying potent methane emissions on a global scale (McKibben 2016a). In the run up to the 2016 presidential election, the movement mobilized its forces to prevent the catastrophe of a Trump presidency through *actively campaigning for* the other bad option—Hillary Clinton. As 350.org staffer Anna Goldstein (2016) argued, "350 Action is launching something that we've never done before: a national week of action to get out the vote ... Voting for Hillary Clinton is the way we know we can defeat Trump and Trump-ism." The influence of Bernie Sanders's path-breaking democratic socialist campaign injected language about climate action into the Democratic Party Platform (2016), but the Hillary Clinton-led Democratic Party was still woefully lacking. As we wrote this chapter, Donald Trump became president of the United States, resetting Obama's labored path to renewable energy in the precise opposite direction—an extremely dangerous situation.

The locus of governmental climate then shifted to California, whose governor, Jerry Brown, struggled with his own legacy on climate change. Often hailed in the press as a climate hero, like Obama, he too continued to support conventional fossil fuel extraction and water-intensive fracking in a state experiencing the worst drought in its history (Public Policy Institute of California 2016). The movement responded to this with a focus on winning local battles. As of 2017, six counties in California had banned fracking. In oil-producing counties, activists came up against intense resistance from Big Oil, especially in Monterey County, whose oil is dirtier than the Alberta tar sands, where they were outspent thirty to one but won the referendum by 56 to 44 percent (Center for Biological Diversity 2016; Rogers 2016). In the case of Santa Barbara County, an anti-fracking measure failed by an almost two to one margin in the 2014 election as the oil industry outspent it by thirteen to one. Did the intervening two years of climate crisis and activism explain Monterey's success?

The period from 2014 onward has been marked by a blockadia-style series of frontline opposition and direct action against fossil fuel projects. A critical highpoint was the historic coming together of Indigenous peoples to stop the Dakota Access Pipeline (DAPL) on Standing Rock Sioux land, ultimately approved by Trump in early 2017, which was the site of critically transformational alliance building across many divides (Donovan 2016). Other bright spots include Shell cancelling its plans to drill in the Alaskan Arctic in September 2015 (in part due to the spirited "kayaktivism" of the movement in the Pacific Northwest; see Brait 2015). Less known is the May 2014 action taken by Ken Ward and Jay O'Hara in Massachusetts, who anchored a lobster boat in front of a barge carrying 40,000 tons of coal, delaying it for a day (Stephenson 2015). That September the power plant that received the coal announced it would shut down in 2017; meanwhile, the local district attorney dropped the main charges against Ward and O'Hara, saying the decision was "made with our concerns for their children, and the children of Bristol County and beyond in mind" (Smith, H. 2014).

After a five-year struggle on many levels, including mobilizations by 350.org and an unlikely but therefore significant "Cowboy-Indian Alliance" in Nebraska, and direct action tree-sitting in Texas on another part of the project, Obama rejected the northern portion of the Keystone XL Pipeline in November 2015, a decision which Donald Trump promptly reversed. In another surprising development, US Magistrate Judge Thomas Coffin ruled in April 2016 that 21 young people have the right to sue the federal government for causing climate change and thus violating their rights to life, liberty, and property. The case is now in district court, and as with much else of potentially dire consequence, up in the air of the dawning age of Trump (Our Children's Trust 2016). Even in such seemingly out of the way places as Idaho, unlikely partners crossed lines to protect their homes and families from extreme energy development (Grosse 2017).

The ways forward and the work yet to be done

What lessons lie in all these events? How can the climate justice movement move forward on all levels? In short, the climate justice movement is a distinct political force in environmental politics in North America—in relation to governments and the broader climate movement. It should continue to boldly push the envelope to elaborate its goals as it engages actively in climate battles.

Frontlines leading

In the words of Rising Tide North America (2015),

> Climate justice is more than just a goal; it's a practice in the movement against climate chaos. No effort to create a livable future will succeed without the empowerment of marginalized communities and the dismantling of the systems of oppression that keep us divided.

Leadership by people on the front lines of climate change is a core principle of climate justice, well-illustrated at Standing Rock. This protest, and the sense of solidarity that moved so many to join the blockade or send resources, amply demonstrates the rising resistance to fossil fuels and creation of alternative—if fragile—futures in their place. From the pilgrimage by Indigenous activists in the "Protect Our Public Lands Tour" in the summer of 2016 (see Foran 2016d) to Kandi Mossett's powerful speech against fracking at COP21 (2015), to Amanda Polchies kneeling with an eagle feather against Royal Canadian Mounted Police— now an iconic image (Shilling 2013), to the Nez Perce blockade of tar sands equipment in Idaho (Johnson 2013)—the most marginalized communities are organizing, with allies standing together in support.

Intersectionality

Building solidarity within intersectional movements is illustrated by 350.org's increasingly explicit statements on the importance of *social* justice. In 2014, 350.org Strategic Partnership Coordinator Deirdre Smith posted the powerful essay "Why the Climate Movement Must Stand with Ferguson" (2014), linking the police murders of black folks with the militarized racism of the aftermath of Hurricane Katrina. Speaking to the lack of focus on intersectionality in the climate movement, Smith wrote that solidarity "involves climate organizers acknowledging and understanding that our fight is not simply with the carbon in the sky, but with the powers on the ground." Likewise, Black Lives Matter recognizes the interconnections of their struggle, acknowledging the threat of climate change in their platform (The Movement for Black Lives 2016). In the spring of 2016, 350.org issued a statement following the massacre of 49 people at a gay nightclub in Orlando, Florida, whose intensity and import makes it difficult not to quote in full:

> We write this with tears in our eyes. We haven't been able to get them out of our eyes since waking up on Sunday morning and seeing the news about the massacre in Orlando, Florida … You might be asking yourself, why is an organization who focuses on climate change, responding to this horrifying night. *Because we're all connected, as is our fight* … To be honest, intersectionality is complicated … Many of us who've fought alongside you to stop climate change, and to ensure communities have clean air and water, are LGBTQ+. Some of us are also Latinx, as were those targeted in Orlando. And some of us are also Muslim, and are already facing a different sort of backlash as a result of this attack. What we need is to be able to *bring our whole selves in this movement* … For us to come together as a movement, and fight for a just future together, we need to know that you're here with us. (350.org 2016, emphasis added)

In short, the climate justice movement, from the national level of 350.org to the grassroots, has thrown itself into the work of movement building, under the banner of "to change everything, we need everyone." This struggle to root movements for social justice in active support of each other, beyond solidarity, is their common goal and the requirement for building movements that rise to the needed level to meet the deepening crisis.

Creativity, love, and hope

Climate justice involves rational and systemic analysis leading to clear analytical strategy, but it is also powered by something more, something that draws on what it means to be human. Surrounding Shell ships and Kinder Morgan facilities with kayaks in Portland (Brait 2015) and British Colombia (Kotyk 2016), holding divestivals to educate and inspire student divestment activists (Fossil Free UCSB 2016), and cultivating community and holding prayer circles while occupying the route of the Dakota Access Pipeline (LeQuesne 2016)—the climate justice movement is resisting and building power with diverse tactics. Josh Fox, who made *Gasland* (2010) and *Gasland 2* (2013) on the horrors of fracking, followed this with *How to Let Go of the World and Love All the Things Climate Can't Change* (2016), a film about love, dovetailing with the focus of some in the movement on healing, or even the quality of life and provision of a kind of planetary hospice as the human species ends.

Encounters within the global *youth* climate justice movement have helped us see very clearly that the young people who are entering this epic fight in growing numbers will be the ones to change everything by coming together and teaching hope, artivism, and the community that comes from love (Foran 2016a; Foran, Gray, and Grosse 2017; Grosse 2019). And in "The Most Important Thing We Can Do to Fight Climate Change Is Try," Rebecca Solnit reminds us that

> We don't have a map for any of this, which is what all the confident prophecies of a predictable, linear future pretend to offer us. Instead, we have, along with the capacity for effort, a compass called hope: a past that we can see, that we can remember, that can guide us along the unpredictable route, along with our commitment to beings now living and yet to be born, that commitment called love. (2015)

Broader coalitions

Standing Rock has shown that the Indigenous frontline can be everyone's frontline when those participants find ways to work together and confront issues of difference and privilege in good faith. The ubiquitous fight against fossil fuel pipelines, ports, oil bomb trains, megaloads, and fracking increasingly shows this too. Meanwhile, the divestment campaign has brought people together across distinct institutions and life stages—young people and community members of all kinds who are transforming universities, churches, and city governments by not only divesting from fossil fuels, but also reinvesting in communities.

The concept of a "just transition" is central to building coalitions between the climate justice and the labor movement and ultimately, bringing them together in one movement. The organization Iron and Earth is led by oil sands workers encouraging growth of the renewable energy sector and re-integration into a variety of trades. The "green energy revolution" outlined in their *Workers Climate Plan* (2016) calls for building a skilled new workforce capacity, promoting renewable energy manufacturing, and redeploying the workforce in new energy sectors, in line with the Canadian Labour Congress's call for C$90 billion investment over five years to create "One Million Climate Jobs" in renewable energy, green building, energy efficiency, and public transport. The Green New Deal presented by Representative Alexandria Ocasio-Cortez and Senator Ed Markey gives hope to US-based movements working on similar goals.

Conclusion

The climate justice movement needs to keep voicing a simple reality: without the shut-down of the fossil fuel industry in total, by stopping the tar sands entirely, ending the reign of coal and the fracking of gas, there will be no realistic attainment of any significant climate goals. In an era in which both the growth imperative of capitalism *and* some forms of official environmentalism continue to feed the spiral into terrifying climate chaos, there is no room for ambivalence, and even less for lazy acquiescence to government policies and UN agreements, which at best can only reflect the more liberal wing of the One Percent (no longer in power in the United States, further underlining the point). What is called for is a commitment to the kind of bold stance and warrior spirit elicited at Standing Rock. Climate justice activists have learned to see through political, media, and consumer manipulation as well as the pervasive, insidious cultural mechanisms that prevent serious confrontation of corporate power and inaction on the climate.

When not denying the existence of climate change altogether, the Trudeau and Trump governments attempt to contain the climate movement, with Trump issuing provocation after provocation, while Trudeau promotes false solutions to the ecological crisis—new technologies, market mechanisms (individual green consumerism, carbon pricing, and tax schemes), or personal lifestyle changes. Both Trump's criminal negligence and Trudeau's half-measures allow for uninterrupted extraction of gas, coal, and oil—not to mention a deadly growth in global energy and material throughput, gross societal inequalities, and the rampant militarism that is inseparable from ever spiraling ecological devastation. The more mainstream climate movement now must be convinced—or forced—to drop the "solutions" offered by the largest corporations in the world, industry lobbyists, and the global political elite who control the systems and infrastructure of globalized capitalism.

As the climate justice frontlines expand, and their connections are more firmly acknowledged, the creativity of all social justice movements reaches ever closer to realizing the unique role and potential power they collectively possess. Only when new forms of participatory political organization, justice-based economies, and the flowering of the ecological imagination become widespread, will we achieve what climate justice seeks and the planet needs: a democratic, egalitarian, culturally diverse, ecological society on a global scale. North America will play its part in this epic struggle.

PART FIVE

THINKING BEYOND THE HERE AND NOW: ENVISIONING MANY FUTURES

27 | CHANGE IS AND WILL BE LOCAL: ALTERNATIVES FROM SPAIN

Anna Pérez Català

If we are to solve the climate crisis, we have to put people at the center. The current world we live in, shaped by an aggressive neoliberal capitalism, has found the most creative ways of destroying the economy, the people, and the planet. And here we find ourselves, in the midst of a deep crisis that will bury us on so many different fronts: we are dangerously speeding towards a more than 2-degrees-warming world, with the enormous consequences that this has for life on the planet as we know it; we have built an economic system that only serves the elites and has exploited the Global South and impoverished half of the North; and we are basing our social relationships on competition, consumerism, and self-interest.

The world needs to stop for a second, look where we are trending towards, and say "this is enough." We know very well the consequences of the world we are creating. And we find ourselves discussing them over and over in workshops, conferences, or debates. We also know the solutions to it: they are nothing new. In many cases, they are linked to things that our grandparents knew and practiced; in some others, technology has helped us reach further, and renewable energy is growing all around.

Maybe we have to stop talking. Stop debating endlessly about what a new world would look like, and just go out there and create it. That means protesting about the issues of the current world that we profoundly dislike, but also rolling up our sleeves and building a new one with our hands. It is time to demonstrate that we can create a different world that is not ruled by the same laws as the one that we live in. A world that takes care of its people and of the environment.

Taking action

When it comes to taking action, institutions have failed us. Neither the global level, nor the national level have found ways to reverse all issues that threaten humanity, prioritizing profit and power before people and planet. This is why change is coming from municipalities, or people organizing at a very local level.

People are taking back local institutions in many different places around the world. The so-called "cities for change" movement in Spain—municipal governments ruled by grassroots or assemblies—are promoting participative democracy, focusing on sustainable transportation, or promoting affordable housing. But we also know and have learned that change will never be entirely driven by our institutions, and that social pressure is essential for stopping evictions, blocking pipeline construction, or fighting for reproductive rights. Decades of experience of the global movements have taught us the importance of keeping up the mobilization and creating alternatives if we want to build a better world.

Alternatives do exist!

I will focus in this chapter on describing some of the alternatives that are already building and growing, hoping that this will inspire other people to take action elsewhere.

There are alternatives growing everywhere in the world, which focus on a whole range of different aspects of our lives. From promoting renewable energy, building different kinds of cities, promoting a different culture of consumerism: in essence, creating a cooperative economy. Change springs in many aspects of our lives. Fostering such change are people who are actually doing what is necessary: stopping to reflect on this economic and social system and taking real action to change it.

SomEnergia

One of the most famous examples of people building a real alternative in my region (Catalonia) is the energy cooperative "SomEnergia,"[1] which means "We are energy." SomEnergia is a non-profit renewable energy cooperative that produces and commercializes green energy to promote a new model based on 100 percent renewable energy. It has close to 40,000 clients. SomEnergia is a cooperative, so it is based on a new way of understanding economy, making decisions in assemblies of members, and promoting transparency, collective engagement, and responsibility.

Spain is known worldwide by its renewable energy production: we have the know-how and the climatic conditions to actually become a 100 percent renewable powered country. The Spanish government, however, keeps hindering the obvious renewable energy expansion in the country, in order to protect the oligopoly of the big energy companies.

All incentives related to renewable energy previously given by the government were dropped (Solchaga 2016), seriously affecting people's investment in new renewable energy plants, and a new fee was introduced in 2015 to tax those who produce their own energy (Tsagas 2015), which also makes it more difficult to be self-sufficient.

Further, big energy producers are often related to corruption cases, revolving doors with the government, and exploitation cases in the Global South. In Spain, high-ranking politicians have become part of companies in the traditional electricity industry (Gil 2013), such as former Presidents Felipe González (Gas Natural) and José María Aznar (Endesa). This often leads to policies that align with the interests of the old electricity utilities and promotes policies that disregard or tax self-production or renewable energy. These revolving doors between the energy giants and the political class have also produced various corruption cases, including a US$4 billion fraud case (Deign 2015), where companies were overpaid by the government and never returned the money to the citizens as agreed.

Finally, many Spanish energy companies are related to abuses in the Global South. For example, Endesa is known across Latin America for impoverishing communities with the construction of dams (Castro 2014), flooding their livelihoods for unsustainable projects, which never give back to the original inhabitants that lose their land.

Especially due to this political situation in Spain regarding renewable energy and the self-production of energy, the work of SomEnergia is very important and powerful. It offers consumers the opportunity to have an independent energy trader that promotes renewable energy, away from the corruption of the big companies.

SomEnergia has become a powerful symbol, and well known throughout Spain, reaching beyond the "early adopters," who were activists and people already worried about environmental issues. The profits that the cooperative makes are returned by lowering the prices to consumers, and the cooperative's more friendly and personalized way of working with the consumers—more human, more caring—makes it a success in a world full of cold and impersonal companies.

The cooperative is also creating its own renewable energy plants, and is taking small investments from its clients to build more. The money that the consumers invest in generating more renewable energy will be translated in lowering their energy prices, thus finding a way to get around the country's above-mentioned restrictive renewable energy policies.

GranollersenTransició (GenT)

"GranollersenTransició,"[2] which could be translated as Transition Granollers, is part of the Transition Network, and its acronym also means "people." The Transition Network[3] is an organization that aims to inspire, encourage, connect, support, and train communities to create initiatives that build resilience and reduce greenhouse gas emissions. The network is well known worldwide, with a strong presence in the United Kingdom, the United States, and Australia. It has a broad range of locally focused initiatives to achieve its goals to increase self-sufficiency and reduce the potential effects of peak oil, climate destruction, and economic instability.

Granollers is a small town outside Barcelona with 60,000 inhabitants. GranollersenTransició is an organization that, being deeply aware of the energy and ecological crises we face, wants to build a different city that promotes a better future. The organization's work focuses on what is feasible and necessary for the city, and its methods include building consciousness about issues, and creating and connecting alternatives.

GranollersenTransició defines itself as an "umbrella" of transition initiatives. It wants to be a meeting point for all small alternatives that are already happening in the city and its surroundings, in order to find better synergies, and work together towards the same goal. As an organization, it mainly works on communication, preparing events, talks, panels, film screenings, and so on to talk about peak oil and climate change. It also lobbies at a local level, and has helped define and influence the local plan for sustainable energy, which informs the city's actions on renewable energy and climate change.

GranollersenTransició also collaborates with a broad range of organizations that promote changes in the economic and ecologic systems of the city. For example, from GranollersenTransició arose a community of bike users, who lobby for better sustainable transportation systems, better bike lanes and parking and, in general, promote the use of bicycles to go to school, work, or move around the city.

GranollersenTransició also works closely with organizations of consumers and producers of organic products. With these partnerships, GranollersenTransició aims to promote alternative and more sustainable and healthy ways of consuming. It fosters the consumption of locally produced, organic, seasonal, and traditional products—mainly local vegetables and grains—thus invigorating the local sustainable producers.

One of the interesting initiatives around the organization's work with local producers is a seed bank of traditional varieties from the region and the idea of an "orchard museum." The orchard museum is grown on land donated by the city council, and it represents a living picture of the traditional and local food varieties of the area. Students and organizations visit the orchard, where they organize all kinds of activities, including communal lunches. With this, they aim to disseminate the lost knowledge on local varieties of fruits and vegetables, explaining how they are grown and cooked, and promoting a more sustainable and local way of buying and cooking.

GranollersenTransició, as well as many other transition towns, is a living example of how change has to be local and integral, if we want it to be durable and real. The organization now plans to amplify the spread of its local currency, to offer economic alternatives to the current system, and also plans to build a better communications network to further disseminate transition alternatives.

Conclusion: fighting the climate crisis on multiple fronts

These are two examples of what people are doing in their daily lives to push back against climate change—people who stop for a second and focus on building a different world.

In every community, region, and country, we will find similar battles that are being fought in order to preserve a healthy climate and build a better society. And the specifics of the fight depend so much on the local context. Sometimes climate action may not be the main reason driving the interest and actions of the actors involved, but since the climate crisis is rooted in many different aspects of our economy and society, there are many ways to fight against that crisis.

In the end, it does not matter if you prefer to focus on energy issues because you do not want to depend on fossil fuels, if you focus on creating a different food culture because you want to promote a healthier people and environment, or if you focus on sustainable transportation because you want more livable cities. They all need grassroots efforts to push for change, and they all will help with, at least, slowing down climate change.

What this global community of people pushing for change reminds the world is that change will not come from above, and that building grassroots movements, cooperatives, and organizations is what will make it happen. SomEnergia does more every day to fight against climate change and build a more democratic and sustainable society than any status quo-oriented government policy will ever do. And that is the great value of the local work.

We need to be able to continue organizing, to find causes that are at the root of the climate crisis, where we can bring true and empowering solutions, and continue building them collectively.

If that continues to happen, I am sure we will be moving towards greater climate justice.

Notes

1. For more details, see www. somenergia.coop/.

2. For more details, see granollersentransicio.wordpress.com/.

3. For more details, see https://transitionnetwork.org/.

28 | "WE'RE DOING SOMETHING TOGETHER THAT WILL REVERBERATE DOWN OVER TIME": AN INTERVIEW WITH BILL MCKIBBEN

John Foran

Bill McKibben (BM) is well-known to an audience passionate about our climate futures. Journalist, author, teacher, farmer, and activist, his has been a tireless voice in the global struggle for climate justice. John Foran (JF) interviewed him over a series of e-mails between April and June 2017.

JF: The idea, Bill, is for us to have a discussion about the prospects for a just climate future from our current vantage point in the first year of the Trump administration.

Let me get right to the point, then:

Donald Trump is now president of the United States. Right-wing, xenophobic governments or parties have momentum in many other wealthy democracies around the world. Even in other places, such as Canada, where Justin Trudeau, a moderate centrist, is in power, no governments anywhere seem very pressed to take serious action on climate change (and this would have been true, of course, had Hillary Clinton become president here in the US). Meanwhile, we are collectively racing through the carbon budget for a less than two degrees Celsius average warming, let alone 1.5.

How does the present moment look to you in terms of our situation on a warming planet? Are there any positive trends at the nation-state level and in the UNFCCC process? Is there any prospect for a better balance of forces in the next several years?

BM: Well, the problem with climate change is always time. One can find positive trends—indeed, the rapid move to renewable energy is hopeful. But it's always been a mathematical stretch to see it catching up to the physics of climate change and that stretch just got a lot further. Trump can and will slow down some of that momentum precisely when it must be accelerating. The planet itself is making its distress clearer, which means that movements will continue to build—but it is possible to make the argument that this Trumpian interregnum comes at an impossible moment.

JF: What do you mean by "an impossible moment?" Do you mean Trump has now reduced our chances for less than 1.5 or 2 degrees from slim to virtually none?

BM: I think it means he has put a roadblock along the path we really need to be traveling—a path that moves upward at an exponential rate. We're so far behind the physics of climate change that simple linear growth in renewables, say, which will continue, won't be enough.

JF: Beyond that, do you think (as I do) that even we—those of us in the global climate justice movement—have yet to internalize the real, horrific, depth of the crisis?

On this point, my friend and ecosocialist scholar-activist Brad Hornick has observed:

All thinking clearly about climate and political realities can do is change the nature of the struggle.

It's not an easy prospect as it requires heart-wrenching personal and collective existential crisis (questioning meaning in all facets of life and work).

I'll say it now: there is conclusive evidence-based scientific determination of irreversible physical changes that will by necessity cause catastrophic destruction to civilization in the coming decades. Full stop.

We are at the point where we need to acknowledge these truths. It will come now or later—and if it comes later it will hit us much harder, and will mean deterioration in the relevance of certain life/work/political strategies.

BM: Well, I wrote the first book about all of this back in 1989, and it was called *The End of Nature*. That should give you some sense where I stand.

JF: For some time, it's been obvious to me that the only real hope for the planet is the climate justice movement. It has certainly made remarkable strides since I first encountered it at the KlimaForum in Copenhagen during the failed UN climate talks of COP15 in

December 2009, where I heard you speak, along with Naomi Klein and President Mohamed Nasheed of the Maldives.

As one of the founders of 350.org—surely one of the biggest climate justice organizations in the world, and the most radical of any of its size (let's not get hung up on this word "radical"—for me it's positive), how do you assess the current state of play in the movement's fight for global climate justice?

BM: It's turning into what may be the world's first truly global movement, which is good since this may be the first truly global problem. I think one of the strongest trends right now is the way people are linking climate change with the other threats posed by fossil fuels: air pollution, usurping and risky infrastructure like pipelines; absurd concentrations of power à la the Koch Brothers. And I think the leadership from grassroots communities has just been immense.

JF: I'm interested in both of these phenomena. Can you discuss some of the most striking examples of grassroots leadership, both here in the US and around the world?

BM: Around the world Indigenous people are in the lead—in the South Pacific, in Australia, in Latin America, in North America. They are concentrated on geographic locations strategic to fossil fuel expansion, and they have enunciated a strong, consistent demand that this kind of colonization come to an end. Front-line communities, because they feel daily the effects that are for most people still somewhat abstract, are in the lead, from inner city communities to rural folks dealing with the spread of fracking.

JF: And as to the links the climate justice movement is making—there appears to be more reaching out to and by movements like, obviously, Standing Rock, and perhaps less obviously, Black Lives Matter. What has to happen for these connections to strengthen and deepen?

BM: I think some of it will happen naturally, because these are the communities getting creamed by climate change, and by the rising environmental neglect as places like the Environmental Protection Agency cut back. But there needs to be much intentional work to connect—to get behind other people's issues. Let's hope Flint and Standing Rock are exemplars, not exceptions.

JF: Along with many other scholar-activists and climate activists generally, I am aware that despite all the growth, creativity, and deepening of the climate justice movement, it is not enough in this race against extinction (there can be no other terminus under business-as-usual in my view).

   What is it going to take to gather such force? *How* would we do it? What are some (all) of your best ideas for a way forward? Take as long as you want on this one!

BM: I think you're right—great creativity is needed. And we're seeing it in the early stages of the resistance to Trump, from the pink hats to the day without immigrants. But the need to build an overwhelming understanding of our peril and our opportunity still exists. We haven't really internalized our danger yet, and until we do action on the relevant scale remains hard. Maybe some old school teach-ins around the planet! We will see.

JF: How do you think we make greater creativity happen? And how do we make more people more aware more quickly of the climate crisis as a matter of life and death?

BM: There needs to be a certain amount of joy in all of this. Look, we're obviously going to take great losses. But we've been given a stunning opportunity in this generation to make a huge, powerful difference that will last for millennia. And so *we should take courage, knowing we're doing something together that will reverberate down over time.* If we do it right—if we break the back of the fossil fuel industry—then we're building a new paradigm for living lighter on this planet, nothing less than a new aesthetic. So, let's try.

JF: What gives you joy? What spreads joy?

BM: Oh well, I like to fight the bad guys. Some days, when I'm pessimistic about the speed at which the climate system is disintegrating, simply causing trouble for the fossil fuel barons has to be enough.

JF: Do you agree with Naomi Klein that if we can make an impact on the climate crisis, this will change everything? What future do you think we can achieve by 2030?

BM: I think Naomi is right. One way is this: if we can get widespread renewable energy, we'll undercut the source of the biggest, most unequal, most politically disgusting industry on this planet, which is fossil fuel. Wind and sun are like kryptonite to people like the Kochs.

JF: In just the time we started this e-mail exchange, Donald Trump has taken the US out of the Paris Agreement. What does this do to the balance of forces in the epic struggle we find ourselves in the middle of?

BM: It makes it clearer what's underway, and the need to resist.

***

Very few people on earth ever get to say: "I am doing, right now, the most important thing I could possibly be doing." If you'll join this fight, that's what you'll get to say. (Bill McKibben)

29 | CLIMATE JUSTICE MUST BE ANTI-PATRIARCHAL, OR IT WILL NOT BE SYSTEMIC

Majandra Rodriguez Acha

System change, not climate change

Climate justice movements—our coalitions, collectives, and organizations; activists, defenders, and educators—across regions ground our vision in the imperative to build "system change, not climate change." Through this, we recognize that climate change is a product of a system of extraction and exploitation of both nature and people—and that the dichotomy between "nature" and "people" is artificial, itself at the root of a system that categorizes life into the *dominated* and the *dominant*. By centering the "system," we acknowledge that climate change is not the problem, but rather a symptom. Just as a fever is a sign of underlying illness in our body, global warming points to a deep-rooted imbalance in our predominant way of life.

The "system," when we begin to ground it in concrete practices and structures, is how our dominant societies, political institutions, and economic systems organize and operate. In these, individualism, self-interest, logocentrism or the superiority of "rationality," and a binary view of the world predominate. At the center is the possibility and desirability of power and material wealth, a zero-sum game that creates haves and have-nots. Within the current umbrella of neoliberal capitalism, these foundational elements are packaged and delivered as "profit," "competitiveness," "growth," and "progress." In this broader system, the Earth is a pool of material resources to be appropriated by "man" through our labor—as John Locke (2014 [1764]) posited—and the environment is a set of conditions that humans can overcome with technological advancement and force.

As we rationalize and seek mass industrial production, hyper-consumption, and accumulation, we have led ourselves to climate chaos and levels of environmental degradation only comparable to the beginnings of prehistoric periods of extinction. We have forced a separation from what gives us the possibility of life, extracting and polluting in disregard of the balance and interdependence of our ecosystems. In doing so, we are passing the limits of the stability of our planet as we know it. Climate change, as researched by a group of scientists housed by the Stockholm Resilience Center, is but *one* of *nine* planetary boundaries that ensure the possibility of human life on Earth (Steffen et al. 2015). We have surpassed the "safe operating space" for four of these boundaries, including biosphere integrity (biodiversity loss and extinctions), biogeochemical flows (excessive nitrogen and phosphorous from industrial and agricultural use), and of course, climate change. Perhaps what is most alarming is what we don't understand: the full extent of our influence in disrupting the Earth's systems.

In this context of systemic crisis, how do we begin to break down these foundations, to center other ways of relating to each other, to ourselves, and to the "natural world?"

Capitalism and patriarchy are interdependent

We cannot draw a complete picture of the systemic crisis we are living in without centering the structural and historic systems of power and oppression based on race, ethnicity, class, gender, and sexuality, and other roles and aspects of identity, that characterize our societies. That is, how the "system" functions cannot be understood without seeing *who* occupies the role of the "dominant," and who the role of the subaltern on a structural level. The "dominated" are those whose bodies, lives, and dignity are taken as means to the end of accumulation: Indigenous groups who have been colonized, workers in the lower rungs, black and brown "minorities," and *women* who are at the crossroads of multiple kinds of oppression. Understanding the mechanisms by which this plays out, and how these systems of oppression rely on each other to function, is fundamental to get at what "system change, not climate change" means and looks like.

The young women, and young feminists with whom I have worked in local and international gatherings and actions for climate and environmental justice, understand the systemic nature of the current climate crisis very well. That is, to many of us, not only can climate change be truly addressed by modifying our deep-rooted economic, political, and social structures, but also by acknowledging that *the system is capitalist as much as it is patriarchal.* As our lived experiences attest, our struggles as women, and as young women, are not separate from our environmental struggles. This is not only because how we are perceived shapes our experiences and power relations as activists and environmental defenders—seen as vulnerable to harassment, not "important" or worthy to listen to—but also in terms of how the capitalist system and patriarchy are intertwined. The deep-rooted systems that are currently in place bring about *both* environmental degradation and the oppression of women.

Mainstream "gender analyses" of climate change often focus on the disproportionate effect of climate impacts on rural and Indigenous women. But, as the World March of Women (Marcha Mundial das Mulheres 2012) describes, "It is not enough to identify that the impacts of the capitalist system are worse for women. An analysis of how capitalism uses patriarchal structures in its current process of accumulation is needed." This includes how the unpaid labor of women to reproduce and care for life is inherently taken as an indirect "subsidy" by our economic system. Women and female bodies create and care for life, often as a primary activity, or as a second or third "job," without monetary remuneration—that is, in economic dependence on habitually male wage-earners. This is compounded by women's socially, economically, and politically subaltern position which naturalizes these roles so that the system can continue to accumulate on our backs.

In the context of environmental degradation, women's care work shoulders the cost that polluters should pay to address human health impacts, particularly in rural areas. At the same time, child-bearing bodies are particularly vulnerable to chemical and other forms of contamination. The feminization of the countryside around the world further means that women are disproportionately relying on subsistence agriculture, while men migrate to cities for paid labor, meaning that women are most exposed and have the least economic resources to face natural disasters such as droughts and floods, and other climate change impacts. Of course, not all women are affected in the same way. The lives, bodies, territories, and livelihoods of rural and Indigenous women are at the core of these intersections, as well as women with precarious economies, the young and elderly, lesbian and queer women, trans and non-binary, women with disabilities, and those who inhabit other intersections of oppression.

From an intersectional feminist perspective, merely saying that women are the most affected by climate change is not only reductionist, glossing over the complexity of power relations and cultural contexts, but can also lead to superficial understandings of the type of change that is needed. Yes, climate initiatives of all kinds must ensure the meaningful participation of women, the equitable distribution of benefits, and avoid increasing the burden and vulnerability of women and other marginalized groups. However, it is necessary to also go deeper, and recognize, question, and uproot the patriarchal configuration of our societies, and patriarchal power itself.

We must transcend the narrative of women as victims, and avoid the trap of essentialisms that naturalize care work as the responsibility of women, and that ascribe the reproduction of life solely to women's bodies, energy, and time. To accomplish this, we must center the sustainability of life—caring for our own and each other's bodies, selves, and lives—as a social, collective, political, and economic priority, and recognize women's care work as an endeavor that requires strength, courage, and wisdom. Essential to this is breaking down dominant, patriarchal dichotomies that separate and oppose "man" to "woman" with all the attendant essentialized binaries: strength–weakness; modernity (future)–tradition (past); civilization–nature; rationality–emotionality, and many others. Breaking down these dichotomies in our mainstream society is as key to destabilizing patriarchal dominance, as it is to delegitimizing the mindset that distances "humans" from "nature" and serves to justify the depredation of the Earth.

Feminist climate futures

In mid-2017, in Nicaragua, I had the opportunity to briefly work with young women environmental activists and defenders from across Central America. They are powerful, courageous, and wise. They are confronting all kinds of violence and harassment, from organized crime to left-wing governments that employ the same tactics of vigilance, repression, and militarization, in alliance with companies vying to take over their community lands, territories, and water, as any right-wing regime. Their groups and alliances are riddled with lawsuits, part of the global trend of environmental advocates being intimidated and burdened by illegitimate Strategic Lawsuits Against Public Participation (SLAPPs) (Saki 2017). These young women tell stories of how they transitioned from being primarily reproductive rights activists, in contexts such as El Salvador, where there are women in prison for having gone through an abortion after they were raped, to being environmental defenders as well. The imposition of large projects, from hydro-power dams to urban expansion, mining, and large-scale industrial agriculture, on their communities cannot be separated from the other forms of oppression and violence that they experience.

They are part of the same package, and reinforce each other. As these young women have found, their environmental activism upsets political and economic systems and institutions of power, and as a result, they are facing increased levels of gender-based violence and intimidation.

If we are to have a truly systemic change, it cannot be just those who directly live and experience the interdependence of capitalism and patriarchy, who understand it and seek to transform it. In other words, we need feminist and climate justice movements to work together—and *our climate justice movements must be feminist in principle and practice*, whether we adopt the term "feminist" or not.

It is not a coincidence that in many of the environmental activist spaces that I have been a part of, women have been a majority, yet it has been common for leadership and speaking roles to be held by male-identifying people, particularly those with socio-economic, racial, and academic privilege. Acknowledging these dynamics, and actively seeking to build a different way of organizing, is key to building truly just, systemic, and effective alternatives. When we walk into a meeting, when we work together on a campaign, when we are out on the streets, we cannot leave parts of our identity out—we come as our whole selves. Our bodies, voices, roles, and relations are with us in our climate activism. Understanding the role that our identities play in our work, in how we are perceived and how we perceive others, and the need to counter all oppressions, is not "identity politics." As Kimberlé Crenshaw (2015), the black feminist legal scholar who coined the term "intersectionality" states: "intersectionality is not just about identities but about the institutions that use identity to exclude and privilege. The better we understand how identities and power work together from one context to another, the less likely our movements for change are to fracture."

To bring about a "system change," we must center the voices, dignity, and right to self-determination of those whom the system uses to keep its cogs turning, including but not limited to, working-class women, and Indigenous and rural women. Our movements must defend their right, our right, to live free from violence and exploitation. Their participation in how we make decisions and how we build climate justice movements is not something to be "allowed" or "accommodated for," but a right that women have won across contexts. Those in privileged positions, and who have historically occupied leadership and visibility, must learn to step aside and out of the spotlight. Ultimately, we must build horizontal structures of operation, take down the spotlight, and build strongly participatory movements.

Feminism, in the understanding of the ecofeminist groups of which I am a part, centers justice, and fights against all forms of domination. We question artificial binaries, and recover the value and power of our emotional selves, as well as our deep interconnectedness with the Earth and all life. This entails unlearning narratives of "human" dominance over "nature," nurturing humility, and recognizing the complexity of the natural systems that we are a part of and that we cannot pretend to understand and predict. It also entails focusing on building relations of solidarity and community, and grounding our activism in the fertile soil of our diversities.

One concrete way through which we can attempt to unlearn, and build systemic movements, is through feminist popular education. Paulo Freire's popular education seeks the transformation of society through dialogue and self-knowledge, organic and intuitive learning, awareness and the critical appraisal of reality, and the construction of new practices and forms of acting. From a feminist perspective, we set out by recognizing that we are all experts in our own realities and experiences, and emphasize learning to think and act in ways that build our *power* in contradistinction from patriarchal notions of power. It is a power oriented by power to do, power to think, and power to feel with autonomy from the mainstream; a power that is not characterized by the fear of its loss or by the taking of the other's power; a power that doesn't classify, and that doesn't need to destroy others and other existing life forms (Agua y Vida 2013: 19).

Examples of feminist popular education activities include the participatory mapping of our struggles and networks; "un-conference" spaces where participants become facilitators; bodily and other exercises that incorporate our emotional, spiritual, affective, and subjective beings; the construction of critical and systemic thinking; learning in collectivity and through creativity; and the recovery of our voice and vision of the world (Agua y Vida 2013: 19 and 21). Through this, we attempt to highlight the personal as political, make our diversities visible, rebuild our relationships, and legitimize our perspective and knowledge, always respecting our bodies and time in the process (Agua y Vida 2013: 34). It is one concrete means to strengthen our collectives, nurture our vision and ourselves, and build healthy movements that are systemic in focus and action.

In conclusion

Climate chaos and environmental degradation necessitate inequality and domination: no being *wants* to experience these negative impacts.

This is compounded by a sense of superiority and separation between "man" and "nature," which has allowed us to not even *see* these impacts as negative for so long. As the global effects of climate change become clearer and more tangible, however, those in dominant power are doing everything they can think of, from technological fixes and back-up plans to spraying chemicals in our clouds and placing reflective shields in space to fleeing to other planets. They seek to escape the consequences of what we are doing, *without changing what we are doing.*

In this race for superficial "solutions" there will continue to be sacrifice zones, those first and most impacted, and those unable to escape the costs of the current system. Ultimately, this race can only end in short-term, ineffective answers to the root problems of exploitation and domination of people and planet. Will we export this to other planets when we're done with ours? When we call for "system change, not climate change," we must ground that in *who* is most impacted and *who* shoulders the costs of the current system, and build alternatives truly rooted in justice. For many around the world, it is clear that we are living a time of transition on many levels—from the climate and environmental crisis, to economic and political shifts and periods of uncertainty. The question must be *when and how* our current political and economic model will end, and *how we are building the foundations for what is to come, here and now.*

30 | IS *VIVIR BIEN* POSSIBLE? CANDID THOUGHTS ABOUT SYSTEMIC ALTERNATIVES[1]

Pablo Solón

Vivir Bien or *Buen Vivir* is a concept that has passed through many different moments. There isn't a single definition of *Vivir Bien*[2] and there are significant differences of opinion on what the term stands for. There are institutions linked to big business that now speak of *Vivir Bien* but their understanding is very different from how it was conceived more than a decade ago in the fight against neoliberalism. It is a space of debate and controversy in which there is no single absolute truth. Indeed, there are many truths as well as countless lies that are today canonized in the name of *Vivir Bien*.

Three decades ago almost no one in South America talked about this concept. What existed then was the *Aymara suma qamaña* and the *Quechua sumaq kawsay*, which express a set of ideas centered in the systems of knowledge, practice, and organization of the native peoples of the Andes of South America. *Suma qamaña* and *suma qawsay* were living realities of the Andean communities, the subject of studies by anthropologists and Aymara and Quechua intellectuals. These ideas went unnoticed by broad sectors of the left and the workers' organizations, especially in urban areas, right through much of the 20th century. Although retreating ever more under the pressure of modernity and development, *Suma qamaña* and *sumaq kawsay* have continued to exist in Andean communities for centuries alongside similar Indigenous visions and terms such as *Teko Kavi* and *Ñandereko* of the Guaraní, *Shiir Waras* of the Shuar and *Küme Mongen* of the Mapuche.

The concept of *Vivir Bien* or *Buen Vivir* began to be theorized toward the late 20th and early 21st century. Perhaps *suma qamaña* and *sumaq kawsay* would never have given way to *Vivir Bien* without the devastating impact of neoliberalism and the Washington consensus. The failure of Soviet socialism, the absence of alternative paradigms, and the advance of privatization and commodification of nature, inspired a process of relearning the Indigenous practices and visions that had been devalued by capitalist modernity.

This process of revalorization occurred in both theory and practice. The dismissal of tens of thousands of workers through the application of neoliberal measures provoked a change in the class structures of the Andean countries of South America. In the case of Bolivia, the miners, who, for almost a century, were the vanguard of all the social sectors, were relocated. In their place, Indigenous peoples and peasants came to the fore.

The Indigenous struggle in defense of their territories not only generated solidarity but awakened interest in understanding this self-managing vision of their territories. Left and progressive intellectuals who had lost their own utopias after the fall of the Berlin Wall began to take a closer look at what could be learned from these Indigenous cosmovisions. That is how *Vivir Bien* emerged. In reality, the term is incomplete and an insufficient translation of *suma qamaña* or *sumaq kawsay*, which have a more complex set of meanings such as "plentiful life," "sweet life," "harmonious life," "sublime life," "inclusive life," or "to know how to live."

A new phase of *Vivir Bien* began with the arrival of the governments of Evo Morales in Bolivia (2006) and Rafael Correa in Ecuador (2007). Both countries institutionalized the term in their new constitutions and *Vivir Bien* came to be a central part of the official discourse of institutional reforms. The national development plans of both countries incorporated it as a point of reference.

These developments prompted links with other alternative visions such as Thomas Berry's "Earth Jurisprudence," generating the development of new proposals like the rights of Mother Earth and the rights of nature, which had not been present originally in *Vivir Bien*. The impact of *Vivir Bien* was so strong that a set of other systemic alternatives like degrowth, the commons, eco-socialism, and others turned their attention at international level toward this vision.

However, this constitutional triumph of *Vivir Bien* was also the beginning of a new phase of controversies in both Bolivia and Ecuador. Questions began to be asked: Is *Vivir Bien* really being applied? Are we moving toward this objective or have we lost our way? The application of *Vivir Bien* which both governments proclaimed nationally and internationally, led to a redefinition of the concept. So what really is *Vivir Bien*? Is it an alternative vision to extractivism or is it a new form of development, more humane and nature friendly?

In Bolivia and Ecuador alike there now exist different interpretations of what is meant by *Vivir Bien* or *Buen Vivir*. Simplistically, we can say that at present we have an official vision that is passable even for financial institutions like the World Bank and another one that is subversive and rebellious. The positions and differences have become so sharp that today longstanding proponents of *Vivir Bien* in both countries think that the respective governments are not actually practicing it. *Vivir Bien* as a paradigm in both countries is in crisis because it has lost credibility in their societies. However, its essence subsists and still nurtures processes of national and international thinking.

We do not know what the future of *Vivir Bien* will be. Perhaps it will end as mere distractionist rhetoric or as a new form of sustainable development. Today, the governments of Ecuador and Bolivia want the concept to adjust to their practices instead of the other way around. In the attempt to canonize their vision of *Vivir Bien*, they have appropriated it with the complicity of the media and innumerable international institutions. It is important to go to the essence of the concept if we are to advance in its actual implementation.

The core elements

There is no decalogue of *Vivir Bien* or *Buen Vivir*. Any attempt to define it in absolute terms would stifle it. What we can do is to approximate its essence. *Buen Vivir* is not a set of cultural, social, environmental, and economic prescriptions but a complex and dynamic mixture that starts from a philosophical conception of time and space and proceeds toward a cosmovision pertaining to the relation between human beings and nature. The strength of *Vivir Bien* in comparison with other alternatives like the commons, degrowth, ecofeminism, deglobalization, eco-socialism, etc. is in the following elements: (1) its vision of the whole or the *Pacha*; (2) coexisting in multipolarity; (3) the pursuit of equilibrium; (4) the complementarity of diverse subjects; and (5) decolonization.

The whole and the *Pacha*

The point of departure of any systemic alternative transformation is its comprehension of the whole. What is the totality in which the process of transformation operates? Can we carry out a profound change in one country alone? Can we be successful if we focus only on economic, social, and institutional aspects? Is the global capitalist system all of the subject matter or is it part of a larger whole?

For *Vivir Bien*, the whole is the *Pacha*. This Andean concept has often been translated simply as Earth. That is why we speak of *Pachamama* as

Mother Earth. However, *Pacha* is a much broader concept that includes the indissoluble unity of space and time. *Pacha* is the whole in constant movement; it is the cosmos in a permanent state of becoming. *Pacha* refers not only to the world of humans, animals, and plants but also to the world above (*Hanaq Pacha*), inhabited by the sun, the moon, and the stars, and the world below (*Ukhu Pacha*), where the dead and the spirits live. For *Vivir Bien*, all of this is interconnected and the whole makes up a unity.

In this space, the past, present, and future co-exist and interrelate dynamically. The Andean vision of time does not follow Newton's mechanics, which state that time is a coordinate independent of space and a magnitude that is identical for each observer. To the contrary, this cosmovision reminds us of Einstein's famous sentence: "The distinction between the past, present, and future is only a stubbornly persistent illusion." Within the concept of the *Pacha*, the past is always present and is recreated by the future. For *Vivir Bien*, time and space are not lineal but cyclical. The lineal notions of growth and progress are not compatible with that vision. Time advances in the form of a spiral. The future is connected to the past. In any advance, there is a return and any return is an advance. Hence, as the Aymara say, to walk forward we have to have our eyes on the past. This spiral vision of time questions the very essence of the notion of "development," of always advancing toward a higher point, of the search to always be better. This ascendant becoming is a fiction for *Vivir Bien*. Any advance involves turns, nothing is eternal, everything is transformed and is a re-encounter of the past, present, and future.

In the *Pacha*, there is no separation between living beings and inert bodies—all have life. Life can only be explained by the relation between all the parts of the whole. There is no dichotomy between living beings and simple objects. Similarly, there is no separation between human beings and nature. All are part of nature and the *Pacha* as an entirety has life.

According to Josef Estermann, the *Pacha*

> is not a machine or a giant mechanism that organizes itself and moves simply by mechanical laws, as stated by the modern European philosophers, especially Descartes and his followers. *Pacha* is rather a living organism in which all parts are related to one another, in constant interdependence and exchange. The basic principle of any "development" should be, then, life (*kawsay, qamaña, jakaña*) in its totality, not only that of humans or animals and plants, but of the whole *Pacha*. (Estermann 2012)

The objective of human beings is not to control nature but to care for nature as one cares for the mother who has given you life. That is the sense of the expression "Mother Earth." Society cannot be understood in relation to human beings alone; it is a community that has nature and the whole at its center. We are the community of the *Pacha*, the community of an indissoluble whole in a permanent process of cyclical change.

Suma qamaña and *sumaq kawsay* are Pachacentric, not anthropocentric. The recognition and relevance to the whole is the key to *Vivir Bien*. The Andean cosmovision places the principle of "totality" at the core of its existence. Material life is only one aspect and cannot be reduced to the accumulation of things and objects. We have to learn to eat well, dance well, sleep well, drink well, to practice one's beliefs, work for the community, take care of nature, appreciate elders, respect whatever surrounds us, and learn as well how to die—because death is an integral part of the cycle of life. In the Aymara way of thinking, there is no death as understood in the West, in which the body disappears into a hell or a heaven. Here, death is just another moment of life, because one lives anew in the mountains or the depths of the lakes or rivers (Mamani 2011).

In this sense, the whole has a spiritual dimension in which the conceptions of self, of the community, and of nature are based on and linked cyclically in space and time. To live in accordance with the whole means living with emotion, concern, self-understanding, and empathy toward others. This cosmovision has a series of concrete implications: favorable policies are those that take into account the whole and not only some parts, and to act only according to the interests of one part (humans, countries of the North, elites, material accumulation, etc.) will inevitably generate imbalances in the whole.

Coexisting in multipolarity

In the *Vivir Bien* vision, there is a duality in everything since everything has contradictory pairs. Pure good does not exist; good and bad always co-exist. Everything is and is not. The individual and the community are two poles of the same unit. An individual is a person only in as much as he or she works for the common good of his or her community. Without community there is no individual and without singular beings there is no community. A person is not strictly speaking a person without his or her partner. The election of authorities is by twos: man–woman, as a couple. This bipolarity or multipolarity of partners is present in everything. The individual–community polarity is immersed in the humanity–nature polarity. The community is a community not only of humans but also of non-humans.

In the Andean communities, individual private property coexists with communal property. There are differences and tensions between members of a community. To manage those tensions various cultural practices are carried out in order to promote some kind of redistribution. This means, for example, that the wealthiest pay for the fiesta of the entire community or are responsible for other acts or services that benefit everyone.

There are also different practices of collaboration within the community. In the *Mink'a* everyone performs collective labor for the community. In the *Ayni* some members of the community support others and in return the latter repay this with support to the former during the seeding, the harvest, or in some other way. In the Andean communities, the principal milestones are not limited only to the individual or his or her family, but are shared with the entire community. When a child is born, the whole community celebrates. Marriage is not only the union of two persons but the union of two families or communities.

The Indigenous communities worldwide are very diverse. They vary from region to region and country to country. But notwithstanding their differences, they share the sense of responsibility and belonging to their communities. The worst punishment is to be expelled from the community; it is worse than death because it is to lose your membership, your essence, your identity. In contrast to this Indigenous practice, Western societies tend to focus on the individual, on personal success, on the rights of the individual and above all on the protection of one's private property through laws and institutions.

The pursuit of equilibrium

For *Vivir Bien*, the objective is the pursuit of equilibrium among the various elements that make up the whole—a harmony not only between human beings but also between humans and nature, between the material and the spiritual, between knowledge and wisdom, between diverse cultures and between different identities and realities. *Vivir Bien* is not a version of development that is simply more democratic, non-anthropocentric, holistic, or humanizing. This cosmovision has not embraced the notion of progress of the Western civilizations. In opposition to permanent growth, it pursues equilibrium. This equilibrium is not eternal or permanent. Any equilibrium will give rise to new contradictions and disparities that call for new actions to rebalance things. That is the principal source of the movement, of the cyclical change in space-time. The pursuit of harmony between human beings and with Mother Earth is not the search for an idyllic state but the raison d'être of the whole.

This equilibrium is not similar to the stability that capitalism promises to achieve through continuous growth. Stability, just like permanent growth, is an illusion. Sooner or later any growth without limits will produce severe upheavals in the *Pacha*, as we are seeing now in the planet. Equilibrium is always dynamic. *Vivir Bien* is not to achieve a paradise, but to pursue the well-being of everyone, the dynamic and changing equilibrium of the whole. Only by understanding the whole in its multiple components and in its becoming is it possible to contribute to the search for new equilibrium and to live in conformity with *Vivir Bien*.

This essential component of *Vivir Bien* has major implications because not only does it challenge the dominant paradigm of growth but also, it promotes a concrete alternative with the pursuit of equilibrium. A society is vigorous not by its growth but because it contributes to equilibrium both between human beings and with nature. It is fundamental in this process to overcome the concept of human beings as "producers," "conquerors," and "transformers" of nature, and to substitute that with "caretakers," "cultivators," and "mediators" of nature.

The complementarity of diverse subjects

Equilibrium between contraries that inhabit a whole can only be achieved through complementarity. Not by cancelling the other but by complementing it. Complementarity means seeing the differences as part of a whole. The objective is how, between these different parts, some of which are antagonistic, we can complement and complete the totality. Differences and particularity are part of nature and life. What we must do is respect diversity and find ways to articulate experiences, knowledges, and ecosystems.

Capitalism operates under a very different dynamic. According to the logic of capital, what is fundamental is competition to increase efficiency. Whatever restricts or limits competition is negative. Competition will ensure that each industry or country specializes in something in which it can gain. In the end, each will become more efficient at something and will encourage innovation and increase productivity. However, from the perspective of complementarity, competition is negative because some win and others lose, unbalancing the totality. Complementarity seeks optimization through the combination of strengths. The more one works together with the other, the greater is the resilience of each and of all.

Complementarity is not neutrality between opposites but recognition of the possibilities that provide the diversity to balance the whole.

In concrete terms, this means that instead of seeking efficiency through equal rules for unequal groups, industries, or countries, we should promote asymmetrical rules that favor the most disadvantaged so that all can rise. "Knowing how to live" is to practice pluriculturalism, to recognize and learn from difference without arrogance or prejudice. Accepting diversity means that in our world there are other *Buen Vivires* in addition to the Andean version. Those *Buen Vivires* survive in the wisdom, knowledge, and practices of peoples who are pursuing their own identity. *Vivir Bien* is a plural concept, both in the recognition of human pluriculturalism and in the existence of diversity of ecosystems in nature (Gudynas and Acosta 2011). It proposes an intercultural encounter between different cultures. There is no single alternative. There are many, which complement each other in order to make up systemic alternatives.

Vivir Bien is not a utopian regression to the past, but the recognition that in the history of humanity there have been, there are, and there will be other forms of cultural, economic, and social organizations that can contribute to overcoming the present systemic crisis to the extent that they complement each other.

Decolonization

In the vision of *Vivir Bien* there is a continual struggle for decolonization. The Spanish conquest 500 years ago initiated a new cycle. That colonization did not end with the processes of independence and constitution of the republics in the 19th century, but it continues under new forms and structures of domination. To decolonize is to dismantle those political, economic, social, cultural, and mental systems that still rule. Decolonization is a long-term process that does not happen once and forever. We can achieve independence from a foreign power and be more dependent on its economic hegemony. We can conquer a certain economic sovereignty yet continue being culturally subjugated. We can be fully acknowledged in our cultural identity by a new constitution of the state and yet continue to be prisoners of a Western consumerist vision. This is perhaps the most difficult part of the decolonization process: liberating our minds and souls, which have been captured by false and alien concepts.

To build *Vivir Bien* we have to decolonize our territories and our being. The decolonization of territory means self-management and self-determination at all levels. Decolonization of the being is even more complex and includes overcoming many beliefs and values that impede our re-encounter with the *Pacha*. In this context, the first step in *Vivir Bien* is to see with our own eyes, to think by ourselves, and to dream with our own dreams. A key point of departure is to encounter our roots, our identity, our history, and our dignity. To decolonize is to reclaim our life, to recover the horizon. To decolonize is not to return to the past but to put the past in the present, to transform memory as an historical subject. As Rafael Bautista puts it,

> The linear course of time of modern physics is no longer of use to us; that is why we need a revolution in thinking, as part of the change. The past is not what is left behind and the future is not what is coming. The more we are conscious of the past, the greater the possibility of producing the future. The real subject of history is not the past as past but the present, because the present is what always needs a future and a past. (Bautista 2010)

Vivir Bien is a plea to recover the past in order to redeem the future, amplifying the overlooked voices of the communities and Mother Earth (Rivera Cusicanqui 2010: 39–63).

Decolonization means rejecting an unjust *status quo* and recovering our capacity to look deeply so as not to be trapped by colonial categories that limit our imagination. To decolonize is to respond to the injustices that are committed against other beings (human and non-human), to break down the false limits between humanity and the natural world, to say aloud what we think, to overcome the fear of being different, and to restore the dynamic and contradictory equilibrium that has been shattered by a dominant system and way of thinking.

* * *

The experience of this decade shows us clearly that *Vivir Bien* is not possible in a single country in the context of a global economy that is capitalist, productivist, extractivist, patriarchal, and anthropocentric. If this vision is to advance and thrive, a key element is its articulation and complementarity with other similar processes in other countries. This process cannot be limited to the promotion of agreements for integration that do not follow the rules of free trade, nor can it exist merely at the level of states or governments. Analysts have stated that probably one of the biggest shortcomings of the last decade was how some alliances of social and Indigenous movements became too close to some progressive governments and hence had lost some of their ability to develop independently. Looking back, some in the global justice movement in Latin America have reflected that instead of becoming stronger, some were weakened by its inability to articulate its own independent vision of change. Some had confused its utopias with the political plans of the progressive governments and lost its capacity to criticize or to dream beyond that.

If the processes of transformation are to flourish, they need to expand beyond the national borders and into the countries that now colonize the planet in different forms. Without that dissemination to the crucial centers of global power, the processes of change will end up isolating themselves and losing vitality until they have repudiated the very principles and values that once gave birth to them. To that extent the future of *Vivir Bien* largely depends on the recovery, reconstruction and empowerment of other visions that to varying degrees point toward the same objective in the different continents of the planet. *Vivir Bien* is possible only through complementarity with and feedback from other systemic alternatives.

Note

1. This chapter is an abridged and edited version of an essay written by Solón in 2016, translated from the original Spanish by Richard Fidler, and posted on Fidler's website, "Life on the left" at https://lifeonleft.blogspot.com/2016/09/vivir-bien-going-beyond-capitalism.html.

2. In Ecuador it is written "sumak kawsay" and in Bolivia "sumaq kawsay."

31 | THE POLITICS OF CLIMATE CHANGE: ALL CHANGE

Jonathon Porritt

Naomi Klein's *This Changes Everything: Capitalism versus the Climate* (Klein 2014; referred to in the Introduction to this volume) is right: no amount of "tweaking" of today's dominant economic orthodoxy will get us through to a secure, equitable, and genuinely sustainable model of prosperity for the whole of humankind. That doesn't necessarily mean the end of capitalism, but it certainly demands a capitalism transformed.

Somewhat more parochially, the decision by UK voters in June 2016 to exit the EU, and the subsequent election in November 2016 of Donald Trump as the next president of the United States, can also be said to "change everything"—from the point of view of climate politics.

Many of those who mendaciously master-minded the UK's referendum campaign are enthusiastic climate denialists, from Nigel Farage to a host of Tory "grandees." Donald Trump and all his principal supporters are climate denialists to a man.

Trump himself clearly harbors some expectations that he will still be president in 2025—precisely that agonizingly short window of time in which we either succeed in setting ourselves, irreversibly, on the path to a fossil-fuel-free economy, or we reconcile ourselves to living in a world disrupted more and more devastatingly by runaway (and potentially irreversible) climate change.

There's a second, equally compelling reason why those two seismic political shocks have changed everything: the demographic dynamics of addressing climate change will never be the same again. Our conscientious, science-based "call to climate action," studiously evolved over the last 25 years, is now seen by many as a busted flush, an inadvertently elitist and insensitive "grand plan" that left the "left behind" feeling even more ignored and even more patronized than they already were.

Right-wing populists in both the UK and the US took full advantage of that. Posturing as "tribunes of the people," they portrayed middle-class climate activism as yet another attack on the working poor, and on all those whose sense of security, community, and prosperity has been fundamentally undermined over the last 40 years. The fact that this "left-behindness" is primarily a consequence of the self-same neo-liberal economics that has so richly benefitted Trump, and all the rest of the one percent of the one percent elite, matters little in these "post-truth" days.

Either we recognize that new political reality for the fundamental game-changer it is, or we stick to a business plan that we know doesn't work. Let me spell that out more clearly: winning today's "climate war" depends entirely on directly addressing the concerns and the economic misery of tens of millions of people in the rich world, as a preamble to addressing the economic misery of billions of people in the still developing economies of the South.

To be fair, climate campaigners have always recognized this mandate, but the powerful language around the need for a *just transition* to an ultra-low-carbon world has been applied more to the fate of small island states, or to the victims on the front line of climate change in poor countries, than it has to our own hard-pressed populations. What we need now is a total makeover, prioritizing the rollout of solar (for instance) in our poorest communities, giving rural communities a direct stake in renewable energy programs, delivering (finally!) on endless promises to retrofit poor quality housing in both the US and the UK, creating hundreds of thousands of new jobs in those parts of our economies where no other new jobs are ever likely to materialize, and proactively and generously supporting those whose jobs will be lost and whose lives will be devastated as we exit the fossil fuel economy. Is this not the import of the overarching motto of the UN's Sustainable Development Goals: "no-one left behind?"

There are now countless situations where that "just transition" has to be designed and deployed—sooner rather than later, well before it becomes too late to avoid further economic pain and political disaffection. We need to see fully fledged Green New Deals in both the US and the UK; we need transition plans right now, funded by the World Bank, for all those in countries such as India, Indonesia, and South Africa, whose lives depend entirely on the coal industry; we need real action on the ground to help communities in rainforest countries whose livelihoods currently depend on the burning or unsustainable use of those forests—without proper compensation or alternative job prospects, those forests will continue to be destroyed. A decade or more of disputatious theorizing about various REDD (Reducing Emissions from Deforestation and Degradation) strategies must now be translated into a full-on flow of resources on the ground.

This claim may sound unrealistic. But unbeknownst to most climate campaigners, the prospects for protecting the world's remaining forests are about to be transformed. From 2021 onwards, *all* the world's major airlines will be legally bound (by their respective governments' mandate) to ensure that any further growth from that point on will be completely carbon-neutral. What that means, within just a few years, is that tens of billions of dollars will be generated (via carbon levies on all international flights) to help protect forests and other threatened ecosystems.

I'm only too aware that lots of climate campaigners won't like this. They've hated the idea of carbon offsetting from the off, homing in on all the "rubbish offsets" perpetrated in the early years. And some of it was bad! But that has prevented them from looking at all the amazing projects going on today in some of the world's poorest countries, with significant social benefits as well as *genuine* carbon offsets.

I've become more and more critical of this kind of default eco-absolutism on the part of Western NGOs. Invariably, it works against the interests of the world's poor, often putting trees before people and morally superior ideals before practical solutions. But we've now run out of the time for the politics of perfectionism.

And we need to get a whole lot smarter about the notion of "putting a proper price on carbon." Like many others, I've always supported the idea of carbon taxes rather than carbon trading, and am particularly drawn to the idea of using revenues from those taxes as an explicitly redistributive mechanism. In *The World We Made* (Porritt 2013) (written in 2013, but looking back from 2050 at what has to happen to enable the transition to a fair, low-carbon world 20 years from now), I was deluded enough to suggest that a "Cap and Prosper Bill" would be passed in the US in 2017! I may have got the timing a bit wrong, but the principle is still sound.

As I explained it in *The World We Made*, in the US context, any company involved in the business of selling oil, gas, or coal would need to purchase permits for the resulting CO_2 emissions, capped at current levels and then reduced year by year through to 2030. All revenues from the auctioned sale of those permits would be paid into an independent Trust Fund, with the lion's share of those revenues then distributed directly to *every US resident*, as a quarterly dividend, on an equal per capita basis. Very small change for the likes of Donald Trump, but highly significant for those suffering economic hardship—as has been demonstrated year after year with a similar scheme in Alaska. Here's how I characterized the outcome:

> There were many more winners than losers, with dividends for the vast majority of US citizens worth far more than the increased cost of the fossil fuels that they used. In fact, Cap and Prosper has proved to be one of the most successful market-based policies in US history. A lot of that success was down to the Rebuild America programme. This was dedicated not only to infrastructure renewal, but also to the retrofitting of all existing housing and non-domestic buildings to minimise energy consumption. The payback for the US economy in terms of jobs, skills and tax revenues, as well as reduced energy bills and increased efficiency, was astonishing. And the weirdest thing was that it was the Republicans that made it all happen. (Porritt 2013: 50)

Such a scenario now looks pretty forlorn, but what we're talking about here is engineering our own economic feedback loops, so that every step towards the low-carbon economy generates its own economic multipliers for those who have benefitted little, if at all, from the last 40 years of neo-liberal "trickle-down" capitalism. Addressing climate change proactively and purposefully would no longer be seen as a threat to the "left behind," but one of the principal means by which those horrendous gaps in job opportunities, health inequalities, security, and economic well-being are gradually narrowed.

And this isn't just about intra-generational justice—narrowing the gaps amongst those alive today—but about *inter*-generational justice—narrowing the gaps across generations. Many young people in the UK, Europe, and the US today have already come to the conclusion that the likelihood of their lives being better than their parents' lives (once a given in the "implicit contract" between parents and their children) is a boat that sailed long ago. They may not blame us personally for this, but the cumulative debt mounts year by year, even before our collective climate guilt kicks in properly. As it inevitably must over the next few years.

Which was why I projected in *The World We Made* that young people would rise up in 2018, in their millions, the world over, to force aside today's self-serving and climate-blind elites, and build some real momentum towards the genuinely equitable, ultra-low economy of the future. That's one date I did get right! And it's impossible not to be deeply inspired by the rolling program of school strikes going on today around the world.

Pretty much all the most exciting work I do at the moment is with young people—around the time of the EU Referendum here in the UK with an organization called "We Are Europe," through some of Forum for the

Future's most progressive Partners now addressing social justice issues in a much more realistic way, and in the UK's 2017 General Election through an initiative that flowered briefly but gloriously (#SheVotes) to persuade young women to feel empowered enough to cast their vote for the candidate that spoke most powerfully to their values and aspirations. It made a huge impact in just a few weeks.

Without some kind of blistering, no-holds-barred feedback loops, generated primarily by young people within the next few years, we can pretty much guarantee that some of the physical feedback loops we're already feeling the impact of will deprive us forever of the prospect of a genuinely just transition. Nowhere are those feedback effects moving faster than in the Arctic Circle, a politically contested part of the world where Presidents Putin and Trump may finally come to recognize what accelerating climate change really means—not just for their national interests, but for the whole of humankind.

Anyone in any doubt about the underpinning science of what is happening in the Arctic should immediately get hold of a copy of *Farewell to Ice* (Wadhams 2017), a stunning and deeply disturbing "report from the Arctic" by renowned climatologist Peter Wadhams. Wadhams has been tracking the disappearance of summer sea ice in the Arctic over the last 30 years, and compellingly demonstrates that what we're now witnessing is a "Death Spiral," with less and less first-year ice lasting long enough to add to the multi-year ice on which the whole system depends.

Wadhams identifies a further six feedback effects (water vapor; Arctic rivers; black carbon; ocean acidification; wind and waves; and feedback caused by the melting of the Greenland ice sheet in the Arctic), before investigating the hypothesis that all this combined warming could be sufficient to trigger the release of billions of tons of methane captured in the frozen sediments under the seabed. He estimates that at least 55 billion tons (50 billion tonnes) of methane could be released over the next ten years. This one-off "methane pulse" could add the equivalent of 0.6°C to projected temperature increases over the next few decades.

Such predictions may be at the outer limit of what mainstream Arctic scientists subscribe to, but Wadhams is withering in his criticism of those scientists, pointing to "a collective failure of nerve by those whose responsibility it is to speak out and advocate action. It seems to be not just climate change deniers who wish to conceal the Arctic methane threat, but also many Arctic scientists."

Which brings us back full circle to the complex phenomenon of "denialism," and the degree to which the use of that term may itself be a contributory factor in today's populist resurgence. In a period of soul-searching recrimination, there emerged an uncanny similarity between the response to the Brexit vote of those on the progressive left in the UK, and those of the progressive liberal tradition in the US after Trump's victory—and that's to castigate ourselves, amongst other things, for adopting the accusatory language of denialism over the last decade or so. Like so many flagellant penitents, climate campaigners whipped themselves into a frenzy of guilt for having appeared superior, condescending, and insensitive in describing those who deny the now irrefutable science of climate change as "ignorant denialists."

For me, this is just stupid. If anything, we should really be whipping ourselves for the careless use of the word "ignorant," for not distinguishing between those who, *in full command of the science and its potential implications for the future of humankind,* continue to deny that science in public, and those who continue to put their trust in those politicians and commentators simply because they're the ones they've come to trust as their best hope in such cruel and troubled times.

The former (including Donald Trump himself) should be called out, trenchantly and consistently, for what they are: self-serving, ideologically driven, deeply immoral abusers of people's trust. That ideological rationale becomes clearer and clearer by the day. As Naomi Klein so eloquently argues in *This Changes Everything,* addressing the challenge of accelerating climate change demands concerted government action, at all levels of the body politic, from both the regulatory and the fiscal perspective, and it further demands a pooling of sovereignty to secure international cooperation.

This is what has happened in the EU over the last 20 years or so, as each individual nation state has gone along with the idea of the EU negotiating as a bloc, setting and agreeing EU-wide targets. To the backward-looking, narrow-minded nationalists who drove the Brexit campaign back in 2016, this represented an abhorrent erosion of national sovereignty—compounded by the fact that only concerted, cross-EU *government* action can possibly deliver an appropriate response to the existential threat of accelerating climate change. Advocates of "small state politics" and of untrammeled market forces grind their teeth in hatred of such hateful heresies.

But pure ideology (which this is) pays little if any attention to the real needs of ordinary citizens. The lives (and, for that matter, the politics) of the majority of people who continue to question the scientific consensus about climate change, for all sorts of reasons, couldn't be further removed from the willful (I've been tempted elsewhere to use the word "evil") denialism of people like Trump and his self-serving funders in the fossil fuel industries. It makes far more sense to see this majority as *victims of that abuse,* not stupid or ignorant, but distressingly vulnerable to a political con-trick of the worst kind.

To make that distinction, empathetically and not patronizingly, between the knowing denialists of today's neo-liberal elites, and the legion of unknowing victims of denial, is now our biggest challenge. This is why an uncompromising commitment to *a just transition,* in today's rich and poor worlds alike, is now *our* last and best hope.

REFERENCES

350.org. 2016. Orlando. June 17, 2016. http://350.org/orlando/?akid=14359.599245.kVHD7h&rd=1&t=9.

Adger, W.N., Barnett, J., Brown, K., Marshall, N. & O'Brien, K. 2012. Cultural Dimensions of Climate Change Impacts and Adaptation. *Nature Climate Change* 3(2): 112–17.

Adger, W.N., Benjaminsen, T.A., Brown, K. & Svarstad, H. 2001. Advancing a Political Ecology of Global Environmental Discourses. *Development and Change* 32(4): 681–715.

Adger, W.N., Dessai, S., Goulden, M., Hulme, M., Lorenzoni, I., Nelson, D.R., Naess, L.O., Wolf, J. & Wreford, A. 2009. Are There Social Limits to Adaptation to Climate Change? *Climatic Change* 93(3–4): 335–54.

Agarwal, A. & Narain, S. 1991. *Global Warming in an Unequal World: A Case of Environmental Colonialism.* New Delhi: Centre for Science and Environment.

Agence France-Presse. 2016a. Armed Guards at India's Dams as Drought Grips Country. *The Guardian.* May 2, 2016. www.theguardian.com/world/2016/may/02/armed-guards-at-indias-dams-as-drought-grips-country.

——. 2016b. French Parliament Votes to Extend State of Emergency until after 2017 Elections. *The Guardian.* December 13, 2016. www.theguardian.com/world/2016/dec/14/french-parliament-votes-to-extend-state-of-emergency-until-after-2017-elections.

Agua y Vida: Mujeres, Derechos y Ambiente A.C. 2013. *Guía Metodológica para la inclusión de la perspectiva de género en el tema de los bienes comunes.* Chiapas, Mexico: Schenerok, A. & Cacho Niño, N.I. https://observatorio.aguayvida.org.mx/media/uploads/guia_metodologica_1.pdf.

Ahmed, A.U., Huq, S., Nasreen, M. & Hassan, A.W.R. 2015. Climate Change and Disaster Management. Final Report for Sectoral Inputs towards the Formulation of Seventh Five Year Plan (2016–2021). www.plancomm.gov.bd/wp-content/uploads/2015/02/11a_Climate-Change-and-Disaster-Management.pdf.

Akpan, U. & Eboh, M. 2017. Zero Oil Days Clearly before Us – Osinbajo. *The Vanguard.* July 25, 2017. www.vanguardngr.com/2017/07/zero-oil-days-clearly-before-us-osinbajo/amp/.

Alber, G. & Roehr, U. 2006. Climate Protection: What's Gender Got to Do with It? *Women and Environments International Magazine* 70/71, 17–20.

Albizua, A. & Zografos, C. 2014. A Values-based Approach to Vulnerability and Adaptation to Climate Change: Applying Q Methodology in the Ebro Delta, Spain. *Environmental Policy and Governance* 24(6): 405–22.

Allard, C. 2017. Nordic Legislation on Protected Areas: How Does It Affect Sámi Customary Rights. In L. Elenius, C. Allard & C. Sandström (Eds.), *Sámi Customary Rights in Modern Landscapes: Nordic Conservation Regimes in Global Context,* 9–24. London: Routledge.

American Geophysical Union. 2013. Human-induced Climate Change Requires Urgent Action. American Geophysical Union Position Statement. August 2013. http://sciencepolicy.agu.org/files/2013/07/

AGU-Climate-Change-Position-Statement_August-2013.pdf.

Anderson, K. 2012. Climate Change Going beyond Dangerous: Brutal Numbers and Tenuous Hope. In Niclas Hällström (Ed.), *What Next? Climate, Development and Equity*, 16–40. Uppsala: Dag Hammarskjöld Foundation. www.whatnext.org/resources/Publications/Volume-III/Single-articles/wnv3_andersson_144.pdf.

Anderson, R. 2015. Seattleand: The Environmental Disaster Waiting to Happen at Monroe State Prison. *Seattle Weekly*. August 11, 2015. http://archive.seattleweekly.com/news/960079-129/seattleand-the-environmental-disaster-waiting-to.

Angus, I. 2016. *Facing the Anthropocene*. New York: Monthly Review. https://monthlyreview.org/product/facing_the_anthropocene/.

Anjaria, J.S. 2006. A View from Mumbai. *Space and Culture* 9(1): 80–2.

Anshelm, J. & Hultman, M. 2014. A Green Fatwa: Climate Change as a Threat to the Masculinity of Industrial Modernity. *NORMA: International Journal for Masculinity Studies* 9(2): 84–96.

Aquino, M. 2009. Tension Roils Peru after Deadly Amazon Clashes. *Reuters*. June 7, 2009. www.reuters.com/article/us-peru-violence-idUSTRE55463G20090607.

Archer, D. 2009. *The Long Thaw: How Humans Are Changing the Next 100,000 Years of the Earth's Climate*. Princeton, NJ: Princeton University Press.

Arora-Jonsson, S. 2011. Virtue and Vulnerability: Discourses on Women, Gender and Climate Change. *Global Environmental Change* 21(2): 744–51.

——. 2016. Does Resilience Have a Culture? Ecocultures and the Politics of Knowledge Production. *Ecological Economics* 121: 98–107.

——. 2018. Across the Development Divide: Environmental Democracy in a North–South Perspective. In T. Marsden (Ed.), *Sage Handbook of Nature*, 3 vols, 737–60. London: Sage.

Arquero, D., Sen, N. & Keleher, T. 2013. Better Together in the South: Building Movements across Race, Gender, and Sexual Orientation. Applied Research Center, New York. August 6, 2013. www.raceforward.org/research/reports/better-together-bridging-lgbt-racial-justice.

Ashwood, L., Harden, N., Bell, M. & Bland, W. 2014. Linked and Situated: Grounded Knowledge. *Rural Sociology* 79(4): 427–52.

Asmundsson, M. 2016. Samebyn Får Rätt—Vindkraften Stör Renarna. *SVT Nyheter Västerbotten*. www.svt.se/nyheter/lokalt/vasterbotten/sambyn-far-ratt-vindkraften-stor-renarna.

Assadourian, E. 2015. Consequences of Consumerism. In S. Nicholson & P. Wapner (Eds.), *Global Environmental Politics: From Person to Planet*, 97–105. Boulder, CO: Paradigm Publishers.

Atkinson, J. & Dougherty, D.S. 2006. Alternative Media and Social Justice Movements: The Development of a Resistance Performance Paradigm of Audience Analysis. *Western Journal of Communication* 70(1): 64–88.

Atwood, M. 2014. Hope and the "Everything Change." Video-recorded interview as part of the "Imagination and Climate Futures Initiative," Arizona State University. https://vimeo.com/118071435.

Australian Public Service Commission. 2007. Tackling Wicked Problems: A Public Policy Perspective. October 25, 2007. www.apsc.gov.au/publications-and-media/archive/publications-archive/tackling-wicked-problems.

Baber, W.F. & Bartlett, R.V. 2005. *Deliberative Environmental Politics: Democracy and Ecological Rationality*. Cambridge, MA: MIT Press.

——. 2009. *Global Democracy and Sustainable Jurisprudence:*

Deliberative Environmental Law. Cambridge, MA: MIT Press.

—. 2015. *Consensus and Global Environmental Governance: Deliberative Democracy in Nature's Regime.* Cambridge, MA: MIT Press.

Bagemihl, B. 1999. *Biological Exuberance: Animal Homosexuality and Natural Diversity.* New York: St. Martin's Press.

Bali Principles of Climate Justice. 2002. EJNet.org. August 29, 2002. www.ejnet.org/ej/bali.pdf.

Banerjee, N., Song, L. & Hasemyer, D. 2015. Exxon's Own Research Confirmed Fossil Fuels' Role in Global Warming Decades Ago. *Inside Climate News.* September 16, 2015. https://insideclimatenews.org/content/Exxon-The-Road-Not-Taken.

Bangladesh Climate Change Trust (BCCT). 2016. Activities of the Bangladesh Climate Change Trust. Presentation made at the 2016 Quiz Award giving ceremony, Bangladesh Climate Change Trust, Ministry of Environment and Forests. September 4, 2016.

Banks, N., Roy, M. & Hulme, D. 2011. Neglecting the Urban Poor in Bangladesh: Research, Policy and Action in the Context of Climate Change. *Environment and Urbanization* 23(2): 487–502.

Baptiste, A.K. & Rhiney, K. 2016. Climate Justice and the Caribbean: An Introduction. *Geoforum* 73: 17–21.

Bardsley, D. & Rogers, G. 2010. Prioritizing Engagement for Sustainable Adaptation to Climate Change: An Example from Natural Resource Management in South Australia. *Society and Natural Resources* 24(1): 1–17.

Barghouti, O. 2011. *Boycott Divestment Sanctions.* Chicago, IL: Haymarket Press.

Barnett, C. & Land, D. 2007. Geographies of Generosity: Beyond the "Moral Turn." *Geoforum* 38(6): 1065–75.

Bartlett, R.V. 1986. Ecological Rationality: Reason and Environmental Policy. *Environmental Ethics* 8: 221–39.

Bassey, N. 2012. *To Cook a Continent: Destructive Extraction and the Climate Crisis in Africa.* Oxford: Pambazuka Press.

—. 2016. *Oil Politics: Echoes of Ecological Wars.* Montreal: Daraja Press.

Basu, R. & Bazaz, A. 2016. Assessing Climate Change Risks and Contextual Vulnerability in Urban Areas of Semi-Arid India: The Case of Bangalore, CARIAA-ASSAR Working Paper 3. International Development Research Centre, Ottawa, Canada and UK Aid, London, United Kingdom. https://understandrisk.org/wp-content/uploads/Assessing-climate-change-risks-and-contextual-vulnerability-in-urban-areas-of-semi-arid-India.pdf.

Batchelor, S. 1997. *Buddhism without Beliefs.* New York: Riverhead Books.

Bautista, R. 2010. *Hacia una constitución del sentido significativo del "vivir bien."* La Paz, Bolivia: Rincón Ediciones.

Beament, E. & Agerholm, H. 2017. Britain "to Ban All Petrol and Diesel Vehicle Sales by 2040." *The Independent.* July 26, 2017. www.independent.co.uk/news/uk/home-news/uk-petrol-diesel-ban-cars-vehicles-britain-sales-fuels-fossil-government-a7860181.html?amp.

Beeler, C. 2017. Navajo Power Plant Likely to Close, Despite Trump's Promises to Save Coal. *Public Radio International's The World.* June 28, 2017. www.pri.org/stories/2017-06-28/navajo-power-plant-likely-close-despite-trumps-promises-save-coal.

Been, V. 1994. Locally Undesirable Land Uses in Minority Neighborhoods: Disproportionate Siting or Market Dynamics? *The Yale Law Journal* 103(6): 1383–422.

Beeson, M. 2010. The Coming of Environmental Authoritarianism. *Environmental Politics* 19(2): 276–94.

Beldi, L. 2016. Carteret Climate Refugees Seek Home. *ABC News.* August 7, 2016. www.abc.net.au/news/2016-08-07/carteret-climate-refugees-new-home/7693950.

Bettini, G. 2014. Climate Migration as an Adaption Strategy: De-securitizing Climate-induced Migration or Making the Unruly Governable? *Critical Studies on Security* 2(2): 180–95.

Beymer-Farris, B.A. & Bassett, T.J. 2012. The REDD Menace: Resurgent Protectionism in Tanzania's Mangrove Forests. *Global Environmental Change* 22(2): 332–41.

Bhavnani, K.-K. 2015. Climate Justice at COP 21. *Huffington Post.* www.huffingtonpost.com/kumkumbhavnani/climate-justice-at-cop-21_b_8684554.html.

Bhavnani, K.-K., Foran, J., Kurian, P. & Munshi, D. (Eds.). 2016. *Feminist Futures: Reimagining Women, Culture, and Development*, 2nd edition. London: Zed Books.

Biermann, F., Abbott, K., Andresen, S., Bäckstrand, K., Bernstein, S., Betsill, M.M., Bulkeley, H., Cashore, B., Clapp, J., Folke, C., Gupta, A., Gupta, J., Haas, P.M., Jordan, A., Kanie, N., Kluvánková-Oravská, T., Lebel, L., Liverman, D., Meadowcroft, J., Mitchell, R.B., Newell, P., Oberthür, S., Olsson, L., Pattberg, P., Sánchez-Rodríguez, R., Schroeder, H., Underdal, A., Vieira, S.C., Vogel, C., Young, O.R., Brock, A. & Zondervan, R. 2012. Navigating the Anthropocene: Improving Earth System Governance. *Science* 335(6074): 1306–7.

Bluepeace. 2007. Atoll Mangroves Absorbed Lethal Power of Tsunami. www.bluepeacemaldives.org/news2007/atoll_mangroves.htm.

Boeve, M. & Fleishman, B. 2017. Case Study: The Fossil Fuel Divestment Campaign. In Civicus (Ed.), *State of Civil Society Report 2017.* Johannesburg: Civicus. www.civicus.org/documents/reports-and-publications/SOCS/2017/essays/case-study-the-fossil-fuel-divestment-campaign.pdf.

Bond, P. 2012. *Politics of Climate Justice: Paralysis Above, Movement Below.* Durban: University of KwaZulu-Natal Press.

——. 2016. Who Wins from "Climate Apartheid"? African Narratives about the Paris COP21. *New Politics* 15(4). http://newpol.org/content/who-wins-%E2%80%9Cclimate-apartheid%E2%80%9D.

Bonneuil, C. & Fressoz, J.-B. 2016. *The Shock of the Anthropocene: The Earth, History and Us.* Translated by David Fernbach. London: Verso.

Boycott45. 2017. Make Trump Bankrupt Again. www.boycott45.org/.

Brait, E. 2015. Portland's Bridge-Hangers and "Kayaktivists" Claim Win in Shell Protest. *The Guardian.* July 31, 2015. www.theguardian.com/business/2015/jul/31/portland-bridge-shell-protest-kayaktivists-fennica-reaction.

Brännlund, I. & Axelsson, P. 2011. Reindeer Management during Colonization of Sami Lands: A Long-term Perspective of Vulnerability and Adaptation Strategies. *Global Environmental Change* 21: 1095–105.

Brannon Jr., H.R., Daughtry, A.C., Perry, D., Whitaker, W.W. & Williams, M. 1957. Radiocarbon Evidence on the Dilution of Atmospheric and Oceanic Carbon by Carbon from Fossil Fuels. *Eos, Transactions American Geophysical Union* 38(5): 643–50.

Braz, R. & Gilmore, C. 2006. Joining Forces: Prisons and Environmental Justice in Recent California Organizing. *Radical History Review* 96: 95–111.

Brecher, J. 2017. Social Self-defense. January 2017. www.jeremybrecher.

org/social-self-defense-protecting-people-and-planet-against-trump-and-trumpism/.

Bronin, S. 2012. The Promise and Perils of Renewable Energy on Tribal Lands. In S.A. Krakoff & E. Rosser (Eds.). *Tribes, Land, and the Environment*, 103–17. Burlington, VT: Ashgate.

Brune, M. 2016. Orlando. Sierra Club Email. June 19, 2016.

Buchner, B., Mazza, F. & Falzon, J. 2016. Global Climate Finance: An Updated View on 2013 and 2014 Flows. Climate Policy Initiative (CPI). October 2016. http://climatepolicyinitiative. org/wp-content/uploads/2016/10/ Global-Climate-Finance-An-Updated-View-on-2013-and-2014-Flows.pdf.

Buck, H., Gammon, A. & Preston, C. 2014. Gender and Geoengineering. *Hypatia* 29(3): 651–701.

Bulkeley, H., Edwards, G.A.S. & Fuller, S. 2014. Contesting Climate Justice in the City: Examining Politics and Practice in Urban Climate Change Experiments. *Global Environmental Change* 25: 31–40.

Bullard, R. 1990. *Dumping in Dixie: Race, Class, and Environmental Quality.* Boulder, CO: Westview Press.

Bullard, R. & Wright, B. (Eds.). 2009. *Race, Place, and Environmental Justice after Hurricane Katrina: Struggles to Reclaim, Rebuild, and Revitalize New Orleans and the Gulf Coast.* Boulder, CO: Westview Press.

——. 2012. *The Wrong Complexion for Protection: How the Government Response to Disaster Endangers African American Communities.* New York: New York University Press.

Butler J. 1990. *Gender Trouble: Feminism and the Subversion of Identity.* New York: Routledge.

Byravan, S. & Chella Rajan, S. 2012. An Evaluation of India's National Action Plan on Climate Change. Chennai: IFMR – Centre for Development Finance and IIT (Madras).

Callison, C. 2014. *How Climate Change Comes to Matter: The Communal Life of Facts.* Durham, NC: Duke University Press.

Cameron, E.S. 2012. Securing Indigenous Politics: A Critique of the Vulnerability and Adaptation Approach to the Human Dimensions of Climate Change in the Canadian Arctic. *Global Environmental Change* 22(1): 103–14.

Campon-Smith, B. 2016. Justin Trudeau's Liberals Unveil Plan to Price Carbon. *The Star.* October 3, 2016. www.thestar.com/news/ canada/2016/10/03/justin-trudeaus-liberals-unveil-plan-to-price-carbon. html.

Canadell, P. & Raupach, M. 2014. Global Carbon Report: Emissions Will Hit New Heights in 2014. *ECOS Magazine.* September 29, 2014. www. ecosmagazine.com/?paper=EC14226.

Capato, L. Jr., Altomare, E., Narasimhan, S., Coronel, B., Arcara, V., Youfrau, E., Gibson, E., Chang, L., Heffernan, D., Thompson, E.R.H. & Mackey, K. 2016. One Week Ago. 350.org Email. June 19, 2016.

Carbon Brief. 2017. Carbon Countdown: Just Four Years Left of the 1.5C Carbon Budget. April 5, 2017. www. carbonbrief.org/analysis-four-years-left-one-point-five-carbon-budget.

Carrington, D. 2016. Fossil Fuel Divestment Funds Double to $5tn in a Year. *The Guardian.* December 12, 2016. www.theguardian. com/environment/2016/dec/12/ fossil-fuel-divestment-funds-double-5tn-in-a-year.

Carvalho, A. 2005. Representing the Politics of the Greenhouse Effect: Discursive Strategies in the British Media. *Critical Discourse Studies* 2(1): 1–29.

——. 2007. Ideological Cultures and Media Discourses on Scientific Knowledge: Re-reading News on Climate Change. *Public Understanding of Science* 16(2): 223–43.

——. (Ed.). 2011. *As Alterações Climáticas, os Media e os Cidadãos*. Coimbra, Portugal: Grácio Editor.

Carvalho, A., van Wessel, M. & Maeseele, P. 2017. Communication Practices and Political Engagement with Climate Change: A Research Agenda. *Environmental Communication* 11(1): 122–35.

Castro, N. 2014. Indígenas contra Endesa: "No queremos más represas en la zona." *Eldiario.es*. March 13, 2014. www.eldiario.es/desalambre/represas-Endesa-conflictos-America-Latina_0_237626997.html. [In Spanish.]

CBC News. 2016. No Support for National Carbon Plan until Pipeline Progress Made, Notley Warns PM. *CBC News*. October 3, 2016. www.cbc.ca/news/canada/edmonton/no-support-for-national-carbon-plan-until-pipeline-progress-made-notley-warns-pm-1.3789167.

Ceballos, G., Ehrlich, P.R., Barnosky, A., García, A., Pringle, R.M. & Palmer, T.M. 2015. Accelerated Modern Human-induced Species Losses: Entering the Sixth Mass Extinction. *Science Advances* 1(5): 1–5.

Center for Biological Diversity. 2016. Analysis: Monterey County Crude Worse for Climate Than Alberta Tar Sands Oil. Press Release. September 21, 2016. www.biologicaldiversity.org/news/press_releases/2016/monterey-county-crude-09-21-2016.html.

Center for Constitutional Rights. 2009. The Case against Shell. March 24, 2009. https://ccrjustice.org/home/get-involved/tools-resources/fact-sheets-and-faqs/factsheet-case-against-shell.

Center for International Environmental Law (CIEL). 2016. Smoke and Fumes: An Introduction to the Deep History of Oil and Climate Change. Video. www.smokeandfumes.org/fumes.

Centre for Science and Environment. 2011. Richest Indians Emit Less Than Poorest Americans. www.cseindia.org/equitywatch/pdf/richest_poorest_emissions.pdf.

Chakrabarty, D. 2009. The Climate of History: Four Theses. *Critical Inquiry* 35(2): 197–222.

——. 2014. Climate and Capital: On Conjoined Histories. *Critical Inquiry* 41: 14–16.

——. 2016a. The Human Significance of the Anthropocene. In B. Latour (Ed.), *Reset Modernity!*, 189–99. Cambridge, MA: MIT Press.

——. 2016b. Whose Anthropocene? A Response. In T. Lekan & R. Emmett (Eds.), *Whose Anthropocene? Revisiting after Nature: Politics and Practice in Dipesh Chakrabarty's Four Theses*. Special Issue of *RCC Perspectives: Transformations in Environment and Society*, 103–13. Rachel Carson Center, Munich.

Chambers, S. 2003. Deliberative Democratic Theory. *Annual Review of Political Science* 6: 307–26.

Chancel, L. & Piketty, T. 2015. *Carbon and Inequality: From Kyoto to Paris*. Paris: Paris School of Economics.

Christoff, P. & Eckersley, R. 2011. Comparing State Responses. In J.S. Dryzek, R.B. Norgaard & D. Schlosberg (Eds.), *The Oxford Handbook of Climate Change and Society*, 431–48. Oxford: Oxford University Press.

Ciais, P., Gasser, T., Paris, J.D., Caldeira, K., Raupach, M.R., Canadell, J.G., Patwardhan, A., Friedlingstein, P., Piao, S.L. & Gitz, V. 2013. Attributing the Increase in Atmospheric CO_2 to Emitters and Absorbers. *Nature Climate Change* 3(10): 926–30.

Ciplet, D.J., Roberts, T. & Khan, M.R. 2015. *Power in a Warming World: The New Global Politics of Climate Change and the Remaking of Environmental Inequality*. Cambridge, MA: MIT Press.

Clark, N. 2010. Volatile Worlds, Vulnerable Bodies: Confronting Abrupt Climate

Change Theory. *Culture and Society* 27(2–3): 31–53.

——. 2011. *Inhuman Nature: Sociable Life on a Dynamic Planet*. London: Sage.

——. 2014. Geo-politics and the Disaster of the Anthropocene. *Sociological Review* 62(S1): 27–8.

——. 2016. Anthropocene Incitements: Toward a Politics and Ethics of Ex-orbitant Planetarity. In R. Van Munster, and R. Sylvest (Eds.), *Assembling the Planet: The Politics of Globality since 1945*, 126–44. London: Routledge.

Clark, N., Chhotray, V. & Few, R. 2013. Global Justice and Disasters. *Geographical Journal* 179(2): 105–13.

Clark, T. 2015. *Ecocriticism on the Edge: The Anthropocene as a Threshold Concept*. London: Bloomsbury.

Clarke, M.T., Halpern, F. & Clark, T. 2015. Climate Change, Scale, and Literary Criticism: A Conversation. *ariel: A Review of International English Literature* 46(3): 1–22.

Clemençon, R. 2012. Welcome to the Anthropocene: Rio+20 and the Meaning of Sustainable Development. *Journal of Environment and Development* 21(3): 311–38.

Climate Central. 2016. Earth Flirts with a 1.5-Degree Celsius Global Warming Threshold. April 20, 2016. www.scientificamerican.com/article/earth-flirts-with-a-1-5-degree-celsius-global-warming-threshold1/.

Climate Crisis and a Betrayed Generation. Letter to *The Guardian*. March 1, 2019. www.theguardian.com/environment/2019/mar/01/youth-climate-change-strikers-open-letter-to-world-leaders.

Climate Law in Our Hands. 2017. www.climatelawinourhands.org/.

Climate Policy Observer. n.d. Alliance of Small Island States (AOSIS). http://climateobserver.org/country-profiles/alliance-of-small-island-states/.

Collyns, D. & Watts, J. 2017. Peru Floods Kill 67 and Spark Criticism of the Country's Climate Change Preparedness. *The Guardian*. March 17, 2017. www.theguardian.com/world/2017/mar/17/peru-floods-ocean-climate-change.

Color of Change. 2016. Coca-Cola: Stop Sponsoring Donald Trump. www.thepetitionsite.com/takeaction/731/873/029/.

Conrad, S. 2011. A Restorative Environmental Justice for Prison e-Waste Recycling. *Peace Review* 23(3): 348–55.

Cooke, F.M., Nordensvard, J., Saat, G.B., Urban, F. & Siciliano, G. 2017. The Limits of Social Protection: The Case of Hydropower Dams and Indigenous Peoples' Land. *Asia and the Pacific Policy Studies* 4(3): 437–50.

Cotton, M. 2013. Deliberating Intergenerational Environmental Equity: A Pragmatic, Future Studies Approach. *Environmental Values* 22(3): 317–37.

Cox, E. 2016. Indigenous Tar Sands Treaty Could Be Trudeau's Worst Nightmare. *Ricochet*. September 22, 2016. https://ricochet.media/en/1420/indigenous-tar-sands-treaty-could-be-trudeaus-worst-nightmare.

Crenshaw, K. 2015. Why Intersectionality Can't Wait. *The Washington Post*. September 24, 2015. www.washingtonpost.com/news/in-theory/wp/2015/09/24/why-intersectionality-cant-wait/.

Crepelle, A. 2018. The United States First Climate Relocation: Recognition, Relocation, and Indigenous Rights at the Isle de Jean Charles. *Belmont Law Review* 6: 1–40.

Cross, K., Gunster, S., Piotrowski, M. & Daub, S. 2015. New Media and Climate Politics: Civic Engagement and Political Efficacy in a Climate of Reluctant Cynicism. Canadian Centre for Policy Alternatives. September

10, 2015. www.policyalternatives.ca/publications/reports/news-media-and-climate-politics.

CTV Ottawa. 2016. Mass Arrest on Parliament Hill during Pipeline Protest. *CTV Ottawa*. October 24, 2016. http://ottawa.ctvnews.ca/mass-arrest-on-parliament-hill-during-pipeline-protest-1.3129195.

Dardar, T. Jr. 2012. Testimony of Chief Thomas Dardar Jr. Principal Chief of the United Houma Nation. Senate Committee on Indian Affairs, *Oversight Hearing on Environmental Changes on Treaty Rights, Traditional Lifestyles and Tribal Homelands*. Washington, DC: United States Congress.

Dauvergne, P. 2017. *Environmentalism of the Rich*. Cambridge, MA: MIT Press.

Davenport, C. 2016. Diplomats Confront New Threat to Paris Climate Pact: Donald Trump. *The New York Times*. November 18, 2016. www.nytimes.com/2016/11/19/us/politics/trump-climate-change.html.

Davies, R. 2016. Papua New Guinea: Several Dead after Floods and Landslides in Morobe Province. Floodlist. October 26, 2016. http://floodlist.com/asia/papua-new-guinea-floods-ocotber-2016.

Davis, A.Y. 2005. *Abolition Democracy: Beyond Empire, Prisons, and Torture*. New York: Seven Stories Press.

De Cicco, G. 2013. COP18: Between Losing Rights and Gender Balance. Association for Women's Rights in Development (AWID). February 22, 2013. www.awid.org/News-Analysis/Friday-Files/COP18-Between-losing-rights-and-gender-balance.

Deen, T. 2012. U.S. Lifestyle Is Not up for Negotiation. *Inter Press Service*. May 1, 2012. www.ipsnews.net/2012/05/us-lifestyle-is-not-up-for-negotiation/.

Deep South Challenge. 2017. Challenge Website. http://deepsouthchallenge.co.nz/.

Deer, S. & Nagle, M.K. 2017. The Rapidly Increasing Extraction of Oil, and Native Women, in North Dakota. *The Federal Lawyer* 35–7 (April).

Deign, J. 2015. Spain's Largest Electric Utilities Targeted in $4 Billion Fraud Case. *Greentech Media*. September 15, 2015. www.greentechmedia.com/articles/read/utilities-named-in-spains-biggest-ever-fraud-affair.

DeLoughrey, E. 2007. *Routes and Roots: Navigating Caribbean and Pacific Island Literatures*. Manoa, HI: University of Hawai'i Press.

Democracy Now! 2016a. Stopping the Snake: Indigenous Protesters Shut Down Construction of Dakota Access Pipeline. August 18, 2016. www.democracynow.org/2016/8/18/stopping_the_snake_indigenous_protesters_shut.

——. 2016b. *North Dakota: Police Arrest over 100 Water Protectors*. October 24, 2016. www.democracynow.org/2016/10/24/headlines/north_dakota_police_arrest_over_100_water_protectors.

——. 2016c. *NYC: Hundreds Disrupt Citibank Shareholder Meeting to Protest Dakota and Keystone Pipelines*. April 26, 2017. www.democracynow.org/2017/4/26/headlines/nyc_hundreds_disrupt_citibank_shareholder_meeting_to_protest_dakota_keystone_pipelines.

Democratic Party. 2016. 2016 Democratic Party Platform. July 21, 2016. www.demconvention.com/wp-content/uploads/2016/07/Democratic-Party-Platform-7.21.16-no-lines.pdf.

den Elzen, M.G.J., Olivier, J.G.J., Höhne, N. & Janssens-Maenhout, G. 2013. Countries' Contribution to Climate Change: Effect of Accounting for All Greenhouse Gases, Recent Trends, Basic Needs and Technological Progress. *Climatic Change* 121(2): 397–412.

Derrida, J. 1995. *The Gift of Death*. Chicago, IL: University of Chicago Press.

Dewey, J. 1916. *Democracy and Education*. New York: Macmillan.

Dhillon, J. & Estes, N. 2016. Introduction: Standing Rock, #NoDAPL, and Mni Wiconi. *Cultural Anthropology*. December 22, 2016. https://culanth. org/fieldsights/1007-introduction-standing-rock-nodapl-and-mni-wiconi.

Di Chiro, G. 2017. Welcome to the (m)Anthropocene? A Feminist-Environmentalist Critique. In S. MacGregor (Ed.), *The Routledge Handbook of Gender and Environment*, 487–505. London: Routledge.

Diaz, V.M. 2011. Voyaging for Anti-colonial Recovery: Austronesian Seafaring, Archipelagic Rethinking, and the Re-mapping of Indigeneity. *Pacific Asia Inquiry* 2(1): 21–32.

Diprose, R. 2002. *Corporeal Generosity: On Giving with Nietzsche, Merleau-Ponty, and Levinas*. Albany, NY: State University of New York Press.

Djoudi, H., Locatelli, B., Vaast, C., Asher, K., Brockhaus, M. & Basnett Sijaapati, B. 2016. Beyond Dichotomies: Gender and Intersecting Inequalities in Climate Change Studies. *Ambio* 45(Suppl. 3): 248–62.

Dodge, J. 2014. Civil Society Organizations and Deliberative Policy Making: Interpreting Environmental Controversies in the Deliberative System. *Policy Sciences* 47(2): 161–85.

Donovan, L. 2016. NODAPL Day to Draw Thousands against Pipeline. *The Bismarck Tribune*. September 12, 2016. http://bismarcktribune.com/news/state-and-regional/nodapl-day-to-draw-thousands-against-pipeline/article_d1db6bd2-23cb-5b68-8243-15e53c7ad5b6.html.

Dryzek, J.S. 1987. *Rational Ecology: Environment and Political Economy*. Oxford: Basil Blackwell.

——. 1996. *Democracy in Capitalist Times: Ideals, Limits, and Struggles*. New York: Oxford University Press.

——. 2009. Democratization as Deliberative Capacity Building. *Comparative Political Studies* 42(11): 1379–402.

Dryzek, J.S., Norgaard, R.B. & Schlosberg, D. 2013. *Climate-Challenged Society*. New York: Oxford University Press.

Dryzek, J.S. & Stevenson, H. 2011. Global Democracy and Earth System Governance. *Ecological Economics* 70: 1865–74.

Dubash, N.K. & Joseph, N.B. 2015. The Institutionalisation of Climate Policy in India: Designing a Development-focused, Co-benefits Based Approach. Climate Initiative Working Paper. Centre for Policy Research, New Delhi. www.cprindia.org/system/tdf/working_papers/Dubash%20and%20Joseph_The%20Institutionalisation%20of%20Climate%20Policy%20in%20India_2015.pdf?file=1&type=node&id=4376&force=1.

Duncan, C. 2009. Maldives First to Go Carbon Neutral. *The Guardian*. March 15, 2009. www.theguardian.com/environment/2009/mar/15/maldives-president-nasheed-carbon-neutral.

Dunlap, R. & Brulle, R.J. (Eds.). 2015. *Climate Change and Society: Sociological Perspectives*. Oxford: Oxford University Press.

Dunlap, R. & McCright, A. 2013. Organized Climate Change Denial. In J.S. Dryzek, R.B. Norgaard & D. Schlosberg (Eds.), *The Oxford Handbook of Climate Change and Society*, 144–60. Oxford: Oxford University Press.

Durkheim, E. 1982 [1895]. *The Rules of Sociological Method and Selected Texts on Sociology and Its Method*. Edited with an introduction by Stephen Lukes. Translated by W.D. Halls. New York: The Free Press.

Dussais, A.M. 2014. Room for a (Sacred) View? American Indian Tribes Confront Visual Desecration Caused by Wind Energy Projects. *American Indian Law Review* 38(2): 336–420.

Eaton, K. 2017. Tribal Members in Oklahoma Defeat Natural Gas Pipeline. *Indian Country Today*. April 12, 2017. https://indiancountrymedianetwork.com/news/native-news/tribal-members-oklahoma-defeat-natural-gas-pipeline-company/.

Elenius, L., Allard, C. & Sandström, C. 2017. *Indigenous Rights in Modern Landscapes: Nordic Conservation Regimes in Global Context.* London: Routledge.

Ellis, E., Maslin, M., Boivin, N. & Bauer, A. 2016. Involve Social Scientists in Defining the Anthropocene. *Nature* 540(7632): 192–3. www.nature.com/news/involve-social-scientists-in-defining-the-anthropocene-1.21090.

Endres, D. 2009. The Rhetoric of Nuclear Colonialism: Rhetorical Exclusion of American Indian Arguments in the Yucca Mountain Nuclear Waste Siting Decision. *Communication and Critical/Cultural Studies* 6(1): 39–60.

Enloe, C. 2013. *Seriously! Investigating Crashes and Crises as If Women Mattered.* Berkeley, CA: University of California Press.

Environment and Climate Change Canada. 2017. Greenhouse Gas Emissions per Person and per Unit Gross Domestic Product. April 13, 2017. www.ec.gc.ca/indicateurs-indicators/default.asp?lang=en&n=79BA5699-1.

Environmental Protection Agency (EPA). 2014. Greenhouse Gas Emissions from a Typical Passenger Vehicle. www.epa.gov/sites/production/files/2016-02/documents/420f14040a.pdf.

Erftemeijer, P.L.A., Riegl, B., Hoeksema, B.W. & Todd, P.A. 2012. Environmental Impacts of Dredging and Other Sediment Disturbances on Corals: A Review. *Marine Pollution Bulletin* 64(9): 1737–65.

Ergas, C. & York, R. 2012. Women's Status and Carbon Dioxide Emissions: A Quantitative Cross-National Analysis. *Social Science Research* 41, 965–76.

Estermann, J. 2012. Crecimiento cancerígeno *versus* el vivir bien: la concepción andina indígena de un desarrollo sostenible como alternativa al desarrollismo occidental. February 23. www.plataformabuenvivir.com/2011/02/estermann-crecimiento-cancer-buen-vivir/.

Evans, L., Hicks, C., Adger, W., Barnett, J., Perry, A., Fidelman, P. & Tobin, R. 2016. Structural and Psycho-social Limits to Climate Change Adaptation in the Great Barrier Reef Region. *PLoS One* 11(3). https://doi.org/10.1371/journal.pone.0150575.

ExxonMobil. 2015. ExxonMobil Earns $32.5 Billion in 2014; $6.6 Billion during Fourth Quarter. February 2, 2015. http://news.exxonmobil.com/press-release/exxonmobil-earns-325-billion-2014-66-billion-during-fourth-quarter.

Exxonsecrets.org. n.d. ExxonMobil Climate Denial Funding 1998–2014. www.exxonsecrets.org/html/index.php.

Faber, D. 1998. The Struggle for Ecological Democracy and Environmental Justice. In D. Faber (Ed.), *The Struggle for Ecological Democracy: Environment Justice Movements in the United States*, 1–26. New York: The Guilford Press.

Finucane, M., Slovic, P., Mertz, C., Flynn, J. & Satterfield, T. 2000. Gender, Race, and Perceived Risk: The "White Male" Effect. *Health, Risk and Society* 2(2): 159–72.

Flannery, T. 2005. *The Weather Makers: The History and Future Impact of Climate Change.* London: Allen Lane.

Fleming, J. 2017. Excuse Us, While We Fix the Sky: WEIRD Supermen and Climate Engineering. In S. MacGregor &

N. Seymour (Eds.), *Men and Nature: Hegemonic Masculinities and Environmental Change*. Special Issue of *RCC Perspectives: Transformations in Environment and Society* 4: 23–8. doi.org/10.5282/rcc/7979.

Foran, J. 2005. *Taking Power: On the Origins of Third World Revolutions*. Cambridge: Cambridge University Press.

——. 2014. Beyond Insurgency to Radical Social Change: The New Situation. *Studies in Social Justice* 8(1): 5–25.

——. 2016a. Re-imagining Climate Justice: What the World Needs Now Is Love, Hope … and You. In K.-K. Bhavnani, J. Foran, P.A. Kurian & D. Munshi (Eds.), *Feminist Futures: Re-imagining Women, Culture and Development*, 2nd edition, 426–32. London: Zed Books.

——. 2016b. Reimagining Radical Climate Justice. In P. Wapner & H. Elver (Eds.), *Re-imagining Climate Change*, 150–170. New York: Routledge.

——. 2016c. The First Draft of History: Thirty-Four of the Best Pieces on the Paris Agreement at COP 21. https://cloudup.com/cdaFzYn961X.

——. 2016d. The Power of Indigenous Activists at the Summit of the Climate Justice Movement. *Resilience*. August 3, 2016. www.resilience.org/stories/2016-08-03/the-power-of-indigenous-activists-at-the-summit-of-the-climate-justice-movement.

Foran, J., Gray, S. & Grosse, C. 2017. Not Yet the End of the World: Political Cultures of Opposition and Creation in the Global Youth Climate Justice Movement. *Interface* 9(2): 353–79.

Fossil Free UC. 2017. UC Santa Barbara's Chancellor Yang Becomes First UC Chancellor to Call for UC's Divestment from Fossil Fuels. May 11, 2017. www.fossilfreeuc.org/.

Fossil Free UCSB. 2016. Class Walkout and Fossil Free Divestival.

Fossil Free UCSB Event. May 26, 2016. www.facebook.com/events/517870335074246/.

Foucault, M. 2002. *The Order of Things*. New York; London: Routledge.

Fox, E. 2016. Another Tropical Cyclone Heads for Storm-battered Fiji. *Aljazeera*. April 6, 2016. www.aljazeera.com/news/2016/04/tropical-cyclone-heads-storm-battered-fiji-160405115650247.html.

Fox, J. 2016. *How to Let Go of the World and Love All the Things Climate Can't Change*. Documentary. New York: International WOW Company & HBO Documentary Films.

Foxwell-Norton, K. 2017. Australian Independent News Media and Climate Change Reporting: The Case of COP21. In R. Hackett, S. Forde, S. Gunster & K. Foxwell-Norton, *Journalism and Climate Crisis: Public Engagement, Media Alternatives*, 144–66. London: Routledge.

Fraser, N. 1995. From Redistribution to Recognition? Dilemmas of Justice in a "Post-Socialist" Age. *New Left Review* 212: 68–93.

——. 2003. Social Justice in the Age of Identity Politics: Redistribution, Recognition, and Participation. In N. Fraser & A. Honneth, *Redistribution or Recognition? A Political-Philosophical Exchange*, 7–88. London: Verso.

Fraser, N. & Honneth, A. 2003. *Redistribution or Recognition?: A Political-Philosophical Exchange*. London: Verso.

Fresque-Baxter, J. & Armitage, D. 2012. Place Identity and Climate Change Adaptation: A Synthesis and Framework for Understanding. *Wiley Interdisciplinary Reviews: Climate Change* 3(3): 251–66.

Friedman, L. 2017. Hurricane Irma Linked to Climate Change? For Some, a Very "Insensitive" Question. *The New York Times*. September 11, 2017.

www.nytimes.com/2017/09/11/climate/hurricane-irma-climate-change.html.

Fukuyama, F. 2014. *Political Order and Political Decay*. New York: Farrar, Straus and Giroux.

Fuller, R.B. 2008. *Operating Manual for Spaceship Earth*. Estate of R. Buckminster Fuller.

Funes, Y. 2017. Politicians and Environmentalists React after Trump Pulls US from Paris Agreement. *Colorlines*. June 2, 2017. www.colorlines.com/articles/politicians-and-environmentalists-react-after-trump-pulls-us-paris-agreement.

Gaard, G. 2015. Ecofeminism and Climate Change. *Women's Studies International Forum* 49: 20–33.

——. 2017. *Critical Ecofeminism*. Lanham, MD: Lexington Books.

General Economics Division (GED). 2014. Capacity Building Strategy for Climate Mainstreaming: A Strategy for Public Sector Planning Professionals. Dhaka: Planning Commission, Ministry of Planning. www.plancomm.gov.bd/wp-content/uploads/2015/08/Capacity_Building_Strategy.pdf.

Ghosh, A. 2016. *The Great Derangement: Climate Change and the Unthinkable*. Chicago, IL: University of Chicago Press.

Gil, A. 2013. Industria energética y política: una puerta giratoria bien engrasada. *Eldiario.es*. November 3, 2013. www.eldiario.es/economia/electricas-expresidentes-exministros_0_192130890.html.

Gilley, B. 2012. Authoritarian Environmentalism and China's Response to Climate Change. *Environmental Politics* 21(2): 287–307.

Glaeser, E. 2012. *Triumph of the City: How Our Greatest Invention Makes Us Richer, Smarter, Greener, Healthier, and Happier*. New York: Penguin.

Godfrey, M., Sophal, C., Kato, T., Piseth, L.V., Dorina, P., Saravy, T., Savora, T. & Sovannarith, S. 2002. Technical Assistance and Capacity Development in an Aid-dependent Economy: The Experience of Cambodia. *World Development* 30(3): 355–73.

Goldberg, S. 2015. ExxonMobil Gave Millions to Climate-Denying Lawmakers Despite Pledge. *The Guardian*. July 15, 2015. www.theguardian.com/environment/2015/jul/15/exxon-mobil-gave-millions-climate-denying-lawmakers.

Goldstein, A. 2016. 350.org email sent to John Foran. No date.

Gonda, N. 2017. Rural Masculinities in Tension: Barriers to Climate Change Adaptation in Nicaragua. In S. MacGregor & N. Seymour (Eds.), *Men and Nature: Hegemonic Masculinities and Environmental Change*. Special Issue of *RCC Perspectives: Transformations in Environment and Society* 4: 69–76. doi.org/10.5282/rcc/7985.

Gordon, L. 2015. UC Sells Off $200 Million in Coal and Oil Sands Investments. *Los Angeles Times*. September 9, 2015. www.latimes.com/local/education/la-me-ln-uc-coal-20150909-story.html.

Gottlieb, R. & Joshi, A. 2013. *Food Justice*. Cambridge, MA: MIT Press.

Grann, D. 2017. *Killers of the Flower Moon: The Osage Murders and the Birth of the FBI*. New York: Penguin.

Grasswick, H. 2014. Climate Change Science and Responsible Trust: A Situated Approach. *Hypatia* 29(3): 541–57.

Green Climate Fund (GCF). 2017. GCF Portfolio. www.greenclimate.fund/what-we-do/portfolio-dashboard.

Griffin, P. 2017. The Carbon Majors Database. Climate Accountability Institute. http://climateaccountability.org/pdf/CarbonMajorsRpt2017%20Jul17.pdf.

Grinde, D.A. & Johansen, B.E. 1995. *Ecocide of Native America: Environmental*

Destruction of Indian Lands and Peoples. Santa Fe, NM: Clear Light.

Grosse, C.E. 2017. Working across Lines: Resisting Extreme Energy Extraction in Idaho and California. PhD Dissertation. Santa Barbara: University of California.

Grosse, C.E. 2019. Climate Justice Movement Building: Values and Cultures of Creation in Santa Barbara, California. *Social Sciences* 8(79): 1–26. doi: 10.3390/socsci8030079.

Groves, C. 2015. The Bomb in My Backyard, the Serpent in My House: Environmental Justice, Risk, and the Colonisation of Attachment. *Environmental Politics* 24: 853–73.

Gudynas, E. & Acosta, A. 2011. La renovación de la crítica al desarrollo y el buen vivir como alternativa. *Revista internacional de filosofía iberoamericana y teoría social. Utopía y Praxis Latinoamericana* 16(53): 71–83.

Guha, R. 1989. Radical American Environmentalism and Wilderness Preservation: A Third World Critique. *Environmental Ethics* 11(1): 71–83.

Gunaratnam, Y. 2013. *Death and the Migrant: Bodies, Borders and Care.* London: Bloomsbury.

Gunaratnam, Y. & Clark, N. 2012. Deep Race: Climate Change and Planetary Humanism. *Darkmatter* 9(1). www.darkmatter101.org/site/2012/07/02/pre-race-post-race-climate-change-and-planetary-humanism/.

Gunster, S. 2012. Radical Optimism: Expanding Visions of Climate Politics in Alternative Media. In A. Carvalho and T.R. Peterson (Eds.), *Climate Change Politics: Communication and Public Engagement*, 239–67. Amherst, NY: Cambria Press.

Gurney, M. 2013. Whither the "Moral Imperative"? The Focus and Framing of Political Rhetoric in the Climate Change Debate in Australia. In L. Lester & B. Hutchins (Eds.), *Environmental Conflict and the Media*, 187–200. New York: Peter Lang.

Hackett, R.A. 2016. Alternative Media for Global Crisis. *Journal of Alternative and Community Media* 1: 14–16.

Hameed, F. & Ali, M. n.d. An Overview of Coastal Stewardship in the Maldives. United Nations Educational, Scientific and Cultural Organization (UNESCO).

Hamilton, C. 2010. *Requiem for a Species: Why We Resist the Truth about Climate Change.* Oxford: Earthscan.

——. 2017a. *Defiant Earth: The Fate of Humans in the Anthropocene.* Cambridge: Polity.

——. 2017b. The Great Climate Silence: We Are on the Edge of the Abyss but We Ignore It. *The Guardian.* May 4, 2017. www.theguardian.com/environment/2017/may/05/the-great-climate-silence-we-are-on-the-edge-of-the-abyss-but-we-ignore-it.

Hanley, C.J. n.d. Climate Trouble May Be Bubbling Up in Far North. *ABC News.* http://abcnews.go.com/Technology/JustOneThing/story?id=8457650.

Hansen, J. 2009. *Storms of My Grandchildren: The Truth about the Coming Climate Catastrophe and Our Last Chance to Save Humanity.* London: Bloomsbury.

Hansen, J., Kharecha, P., Sato, M., Epstein, P., Hearty, P.J., Hoegh-Guldberg, O., Parmesan, C., Rahmstorf, S., Rockstrom, J., Rohling, E.J., Sachs, J., Smith, P., Steffen, K., von Schuckmann, K. & Zachos, J.C. 2011. The Case for Young People and Nature: A Path to a Healthy, Natural, Prosperous Future. www.columbia.edu/~jeh1/mailings/2011/20110505_CaseForYoungPeople.pdf.

Harari, Y.N. 2015. *Sapiens: A Brief History of Humankind.* New York: HarperCollins.

Haraway, D. 1988. Situated Knowledges: The Science Question in Feminism and the Privilege of Partial Perspective. *Feminist Studies* 14(3): 575–99.

Haraway, D., Ishikawa, N., Gilbert, S.F., Olwig, K., Tsing, A. & Bubandt, N. 2016.

Anthropologists Are Talking: About the Anthropocene. *Ethnos* 81: 535–64.

Hardin, G. 1974. Commentary: Living on a Lifeboat. *BioScience* 24(10): 561–8.

Harris Interactive. 2009. LGBT Americans Think, Act, Vote More Green Than Others. *Business Wire* (New York). October 26, 2009. www.businesswire.com/news/home/20091026005727/en/LGBT-Americans-Act-Vote-Green.

Harvard Business Review (HBR). 2015. Tesla's Not as Disruptive as You Might Think. *Harvard Business Review*. May. https://hbr.org/product/tesla-s-not-as-disruptive-as-you-might-think/F1505A-HCB-ENG?referral=03069.

He, J. 2015. The Human Rights Dimension of Global Climate Change. *Human Rights* 2015(5). [In Chinese.]

Hedges, C. 2014. The Myth of Human Progress and the Collapse of Complex Societies. January 27, 2014. www.truthdig.com/articles/the-myth-of-human-progress-and-the-collapse-of-complex-societies/.

——. 2016. Defying the Politics of Fear. Truthdig. November 7, 2016. www.truthdig.com/articles/defying-the-politics-of-fear/.

Heede, R. 2014. Tracing Anthropogenic Carbon Dioxide and Methane Emissions to Fossil Fuel and Cement Producers, 1854–2010. *Climatic Change* 122: 229–41.

Hellmark, M. 2016. Sveriges Ömma Punkt: Samer Och Gruvor. *Sveriges Natur*. www.sverigesnatur.org/aktuellt/samer-kampar-mot-gruvor/.

Henderson, J.C. 2008. The Politics of Tourism: A Perspective from the Maldives. *Tourismos* 3(1): 99–115.

Herman, R. 2016. Traditional Knowledge in a Time of Crisis: Climate Change, Culture and Communication. *Sustainability Science* 11(1): 163–76.

Hermann, V. 2015. Climate Change, Arctic Aesthetics and Indigenous Agency in the Age of the Anthropocene. *The Yearbook of Polar Law* 7: 375–409.

Heynen, N. 2016. Urban Political Ecology II: The Abolitionist Century. *Progress in Human Geography* 40(6): 839–45.

Hildebrandt, T. 2013. *Social Organizations and the Authoritarian State in China*. New York: Cambridge University Press.

Hilder, T. 2015. *Sámi Musical Performance and the Politics of Indigeneity in Northern Europe*. Lanham, MD: Rowman & Littlefield.

Hiltner, K. 2011. *What Else Is Pastoral? Renaissance Literature and the Environment*. Ithaca, NY: Cornell University Press.

Hilton, I. & Kerr, O. 2016. The Paris Agreement: China's "New Normal" Role in International Climate Negotiations. *Climate Policy* 17: 48–58.

Hislop, M. 2016. How the Trudeau Government Tore Up the Rulebook on Pipelines. *Canadian Business*. July 21, 2016. www.canadianbusiness.com/economy/how-the-trudeau-government-tore-up-the-rulebook-on-pipelines/.

Hoffmeister, V., Averill, M. & Huq, S. 2016. The Role of Universities in Capacity Building under the Paris Agreement. ICCCAD Policy Brief. July 4, 2016. www.icccad.net/wp-content/uploads/2016/08/Capacity-Building-Policy-Brief-July-4.pdf.

Holderness, T. & Turpin, E. 2016. How Tweeting about Floods Became a Civic Duty in Jakarta. *The Guardian*. January 25, 2016. http://goo.gl/I1uOVt.

Holz, C., Kartha, S. & Athanasiou, T. 2017. Fairly Sharing 1.5: National Fair Shares of a 1.5°C-Compliant Global Mitigation Effort. *International Environmental Agreements: Politics, Law and Economics* 13(3): 1–18. https://doi.org/10.1007/s10784-017-9371-z.

Honig, B. 1994. Difference, Dilemmas, and the Politics of Home. *Social Research* 61 (3): 563–97.

Hopke, J.E. 2012. Water Gives Life: Framing an Environmental Justice

Movement in the Mainstream and Alternative Salvadoran Press. *Environmental Communication: A Journal of Nature and Culture* 6(3): 365–82.

Horn, S. 2017. Trump Attorney Sues Greenpeace over Dakota Access in $300 Million Racketeering Case. *DeSmogBlog.* August 22, 2017. www.desmogblog.com/2017/08/22/dakota-access-trump-greenpeace-racketeering?utm_source=dsb%20 newsletter.

Houser, S., Teller, V., MacCracken, M., Gough, R. & Spears, P. 2001. Potential Consequences of Climate Variability and Vhange for Native Peoples and Homelands. In National Assessment Synthesis Team (Ed.), *Climate Change Impacts on the United States*, 351–78. Washington, DC: US Global Change Research Program.

Hufbauer, G.C. & Kim, J. 2009. Climate Policy Options and the World Trade Organization. Economics Discussion Papers. *Economics: The Open-Access, Open-Assessment E-Journal* 3 (2009–29): 1–15. http://dx.doi.org/10.5018/economics-ejournal.ja.2009-29.

Hughey, K.F.D. & Becken, S. 2014. Understanding Climate Coping as a Basis for Strategic Climate Change Adaptation: The Case of Queenstown-Lake Wanaka, New Zealand. *Global Environmental Change* 27: 168–79.

Hulme, M. 2008. The Conquering of Climate: Discourses of Fear and Their Dissolution. *The Geographical Journal* 174(1): 5–16.

Hultman, M. 2017. Exploring Industrial, Eco-modern, and Ecological Masculinities. In S. MacGregor (Ed.), *The Routledge Handbook of Gender and Environment*, 239–52. London: Routledge.

Human Rights Clinic. 2015. Reckless Indifference: Deadly Heat in Texas Prisons. Human Rights Clinic, University of Texas School of Law, Austin, TX. https://law.utexas.edu/wp-content/uploads/sites/11/2015/04/2015-HRC-USA-Reckless-Indifference-Report.pdf.

Human Rights Council. 2015. Summary prepared by the Office of the United Nations High Commissioner for Human Rights. Working Group on the Universal Periodic Review Twenty-first Session. United Nations General Assembly 19–30 January 2015. A/HRC/WG.6/21/SWE/3.

Huntington, S. 1965. Political Development and Political Decay. *World Politics* 17(3): 386–430.

Huq, S. 2016. Why Universities, Not Consultants, Should Benefit from Climate Funds. *Climate Home.* May 17, 2016. www.climatechangenews.com/2016/05/17/why-universities-not-consultants-should-benefit-from-climate-funds/.

Huq, S. & Nasir, N. 2016. Stop Sending Climate Consultants to Poor Countries: Invest in Universities Instead. *The Conversation.* October 3, 2016. https://theconversation.com/stop-sending-climate-consultants-to-poor-countries-invest-in-universities-instead-65135.

IFAD and The Global Mechanism of the United Nations Convention to Combat Desertification. 2009. Climate Change Impacts: Pacific Islands. December 14, 2012. www.ifad.org/documents/10180/9054c140-a03c-4c6e-ae9e-ab9d7972dd19.

Incropera, F.P. 2016. *Climate Change: A Wicked Problem. Complexity and Uncertainty at the Intersection of Science, Economics, Politics, and Human Behavior.* New York: Cambridge University Press.

Inglehart, R. & Norris, P. 2016. Trump, Brexit, and the Rise of Populism: Economic Have-nots and Cultural Backlash. Faculty Research Working Paper Series. John F. Kennedy School of Government, Harvard University.

International Monetary Fund (IMF). 2015. IMF Survey: Counting the Cost of Energy Subsidies. July 17, 2015. www.imf.org/en/News/Articles/2015/09/28/04/53/sonewo70215a.

International Organization for Migration. 2016. Floods and Landslides Follow Drought in PNG Highlands. Reliefweb. March 31, 2016. http://reliefweb.int/report/papua-new-guinea/floods-and-landslides-follow-drought-png-highlands.

Iron and Earth. 2016. Workers Climate Plan. www.workersclimateplan.ca/.

ITUC-Africa. 2015. Concept Note for an African Trade Union Symposium on the Impact of Climate Change on Jobs in Africa: A Trade Union Response. November 23–24.

Jabeen, H., Johnson, C. & Allen, A. 2010. Built-in Resilience: Learning from Grassroots Coping Strategies for Climate Variability. *Environment and Urbanization* 22(2): 415–31.

Jacob, C., McDaniels, T. & Hinch, S. 2010. Indigenous Culture and Adaptation to Climate Change: Sockeye Salmon and the St'át'imc People. *Mitigation and Adaptation Strategies for Global Change* 15(8): 859–76.

Jain, G., Singh, C., Coelho, K. & Malladi, T. 2017. *Long-term Implications of Humanitarian Responses: The Case of Chennai.* London: International Institute for Environment and Development.

James, J. (Ed.). 2003. *Imprisoned Intellectuals: America's Political Prisoners Write on Life, Liberation, and Rebellion.* Lanham, MD: Rowman & Littlefield.

Jantarasami, L.C., Novak, R., Delgado, R., Marino, E., McNeeley, S., Narducci, C., Raymond-Yakoubian, J., Singletary, L. & Whyte, K.P. 2018. Tribes and Indigenous Peoples. In D.R. Reidmiller, C.W. Avery, D.R. Easterling, K.E. Kunkel, K.L.M. Lewis, T.K. Maycock & B.C. Stewart (Eds.), *Impacts, Risks, and Adaptation in the United States: Fourth National Climate Assessment, Volume II,* 572–603. Washington, DC: US Global Change Research Program.

Jasanoff, S. 2010. A New Climate for Society. *Theory, Culture and Society* 27(2–3): 233–53.

Johansson, T. 2010. Fullbordat Faktum: En Studie av Förankringsprocesser i Samband med Vindkraftsetableringar i Piteå Kommun. www.pitea.se/contentassets/24fb8688eca2 44f3bb6c493a77250330/fullbordat-faktum--en-demokratistudie.pdf.

Johnson, K. 2013. Fight over Energy Finds a New Front in a Corner of Idaho. *The New York Times.* September 25, 2013. www.nytimes.com/2013/09/26/us/fight-over-energy-finds-a-new-front-in-a-corner-of-idaho.html.

Johnsson-Latham, G. 2007. A Study on Gender Equality as a Prerequisite for Sustainable Development. Report to the Environment Advisory Council. Ministry of the Environment, Stockholm, Sweden. www.uft.oekologie.uni-bremen.de/hartmutkoehler_fuer_studierende/MEC/09-MEC-reading/gender%202007%20EAC%20rapport_engelska.pdf.

Johnston, I. 2017. World Could Put Carbon Tax on US Imports if Donald Trump Ditches Paris Agreement, Says Expert. *The Independent.* May 23, 2017. www.independent.co.uk/environment/carbon-tax-us-imports-world-donald-trump-paris-agreement-climate-change-global-warming-a7751531.html.

Kahan, D.M., Peters, E., Wittlin, M., Slovic, P., Ouellette, L., Braman, D. & Mandel, G. 2012. The Polarizing Impact of Science Literacy and Numeracy on Perceived Climate Change Risks. *Nature Climate Change* 2(10): 732–5.

Kalaugher, E., Bornman, J., Clark, A. & Beukes, P. 2013. An Integrated Biophysical and Socio-economic

Framework for Analysis of Climate Change Adaptation Strategies: The Case of a New Zealand Dairy Farming System. *Environmental Modelling and Software* 39: 176–87.

Karpowitz, C., Raphael, C. & Hammond, A.S. 2009. Deliberative Democracy and Inequality: Two Cheers for Enclave Deliberation among the Disempowered. *Politics and Society* 37(4): 576–615.

Kasrils, R. 2015. Whither Palestine? SAHO South African History Online. May 19, 2015. www.sahistory.org.za/archive/whither-palestine-ronnie-kasrils-19-may-2015-london.

Keijzer, N. 2013. *Unfinished Agenda or Overtaken by Events? Applying Aid and Development Effectiveness Principles to Capacity Development Support.* Bonn: Deutsches Institut für Entwicklungspolitik.

Keijzer, N. & Janus, H. 2014. Linking Results-based Aid and Capacity Development Support Conceptual and Practical Challenges. Discussion Paper 25/2014. German Development Institute, Bonn.

Kelman, I. 2014. No Change from Climate Change: Vulnerability and Small Island Developing States. *The Geographical Journal* 180(2): 120–9.

Kench, P.S. & Brander, R.W. 2006. Response of Reef Island Shorelines to Seasonal Climate Oscillations: South Maalhosmadulu Atoll, Maldives. *Journal of Geophysical Research: Earth Surface* 111(F1): F01001.

Kench, P.S. & Ford, M.R. 2015. Multi-decadal Shoreline Changes in Response to Sea Level Rise in the Marshall Islands. *Anthropocene* 11: 14–23.

Kentish, B. 2016. Nicolas Sarkozy Promises to Hit America with a Carbon Tax if Donald Trump Rips Up Landmark Paris Climate Deal. *The Independent.* November 15, 2016. www.independent.co.uk/news/world/europe/ donald-trump-us-carbon-tax-nicolas-sarkozy-global-warming-paris-climate-deal-a7418301.html.

Khan, M., Sagar, A., Huq, S. & Thiam, P.K. 2016. Capacity Building under the Paris Agreement. European Capacity Building Initiative. www.eurocapacity.org/downloads/Capacity_Building_under_Paris_Agreement_2016.pdf.

King, E. 2013. Pacific Leaders Adopt Majuro Declaration for Climate Change. *The Guardian.* September 5, 2013. www.theguardian.com/environment/2013/sep/05/pacific-leaders-majuro-declaration-climate-change.

Kirilenko, A.P. & Stepchenkova, S.O. 2014. Public Microblogging on Climate Change: One Year of Twitter Worldwide. *Global Environmental Change* 26: 171–82.

Klein, N. 2014a. *This Changes Everything: Capitalism vs. the Climate.* New York: Simon & Schuster.

——. 2014b. Why #BlackLivesMatter Should Transform the Climate Debate. *The Nation.* December 12, 2014. www.thenation.com/article/what-does-blacklivesmatter-have-do-climate-change/.

——. 2016a. Canada's Founding Myths Hold Us Back from Addressing Climate Change. *The Globe and Mail.* September 23, 2016. www.theglobeandmail.com/news/national/canadas-founding-myths-hold-us-back-from-addressing-climate-change/article32022126/.

——. 2016b. Donald Trump Isn't the End of the World, but Climate Change May Be. November 10, 2016. http://sydneypeacefoundation.org.au/naomi-klein-donald-trump-isnt-the-end-of-the-world-but-climate-change-may-be/.

——. 2017. Will Trump's Slo-mo Walkaway World in Flames behind Him Finally Provoke Consequences for Planetary Arson? *The Intercept.* June 1, 2017.

https://theintercept.com/2017/06/01/
will-trumps-slow-mo-walkaway-world-
in-flames-behind-him-finally-provoke-
consequences-for-planetary-arson/.

Klinsky, S., Dowlatabadi, H. &
Mcdaniels, T. 2012. Comparing
Public Rationales for Justice Trade-
offs in Mitigation and Adaptation
Climate Policy Dilemmas. *Global
Environmental Change* 22(4): 862–76.

Knight, N. 2016. "Betrayal": Trudeau
Approves Kinder Morgan Pipeline
Expansion. *Common Dreams.*
November 30, 2016. www.
commondreams.org/news/2016/11/30/
betrayal-trudeau-approves-kinder-
morgan-pipeline-expansion.

Kothari, U. 2014. Political Discourse
of Climate Change and Migration:
Resettlement Policies in the Maldives.
The Geographical Journal 180(2):
130–40.

Kotlikoff, L. 2017. Will "Boycott
America" Follow Trump's Paris
Accord Withdrawal? *Forbes.* June
1, 2017. www.forbes.com/sites/
kotlikoff/2017/06/01/will-boycott-
america-follow-trumps-paris-accord-
withdrawal/#4524daad2dda.

Kotyk, A. 2016. Climate Activists
Surround Kinder Morgan Facility by
Land and Sea. *Rabble.* May 16, 2016.
http://rabble.ca/news/2016/05/
climate-activists-surround-kinder-
morgan-facility-land-and-sea.

Kurian, P. 2017. Feminist Futures in the
Anthropocene: Sustainable Citizenship
and the Challenges of Climate Change
and Social Justice. *Women's Studies
Journal* 31(1): 104–7.

Leap Manifesto. 2015. https://
leapmanifesto.org/en/the-leap-
manifesto/.

Leiserowitz, A., Maibach, E., Roser-
Renouf, C., Rosenthal, S. & Cutler, M.
2016. *Politics and Global Warming,
November 2016.* Yale University
and George Mason University.
New Haven, CT: Yale Program on

Climate Change Communication.
http://climatecommunication.yale.
edu/wp-content/uploads/2016/12/
Global-Warming-Policy-Politics-
November-2016.pdf.

Lelu Island. 2016. Stop Pacific
NorthWest LNG/Petronas on
Lelu Island. www.facebook.com/
Stop-Pacific-NorthWest-LNGPetronas-
on-Lelu-Island-949045868451061/.

Lenferna, A. 2017. Trump Is Withdrawing
from the Paris Climate Agreement:
Is It Time to Boycott America? *Daily
Kos.* June 1, 2017. www.dailykos.com/
stories/2017/6/1/1668071/-Trump-
is-Withdrawing-from-the-Paris-
Climate-Agreement-Is-it-Time-to-
Boycott-America.

Leonard, S., Parsons, M., Olawsky, K. &
Kofod, F. 2013. The Role of Culture
and Traditional Knowledge in Climate
Change Adaptation: Insights from
East Kimberley, Australia. *Global
Environmental Change* 23(3): 623–32.

LeQuesne, T. 2016. Winter Is
Coming: Standing Rock Digs in
for the Long Haul. *Resilience.*
September 22, 2016. www.
resilience.org/stories/2016-09-22/
winter-is-coming-standing-rock-digs-
in-for-the-long-haul.

Lerner, S. 2006. *Diamond: A Struggle for
Environmental Justice in Louisiana's
Chemical Corridor.* Cambridge, MA:
MIT Press.

Lever-Tracy, C. (Ed.). 2010. *Routledge
Handbook of Climate Change and
Society.* London: Routledge.

Levin, K., Cashore, B., Bernstein, S.
& Auld, G. 2012. Overcoming the
Tragedy of Super Wicked Problems:
Constraining Our Future Selves to
Ameliorate Global Climate Change.
Policy Sciences 45: 123–52. https://
munkschool.utoronto.ca/egl/
files/2015/01/Overcoming-the-
tradegy-of-super-wicked-problems.pdf.

Lewis, J. 1990. Conference on Sea-Level
Rise. *AODRO Newsletter* 8(2).

Li, J., Cai, Q., Ma, C., Wang, J., Zhou, Z. & Wang, T. 2016. China's Climate Change Policy and Market Prospect. *Energy of China* 38(1): 5–21. [In Chinese.]

Lingis, A. 1998. Translator's Introduction. In E. Levinas (Ed.), *Otherwise Than Being or Beyond Essence*, xvii–xlv. London: Kluwer Academic Publishing.

Locke, J. 2014 [1764]. Of Property. In *The Second Treatise of Civil Government*, 6th edition. South Australia: eBooks@ Adelaide, The University of Adelaide. https://ebooks.adelaide.edu.au/l/locke/john/l81s/index.html.

Lowe, T., Brown, K., Dessai, S., Doria, M.F., Haynes, K. & Vincent, K. 2006. Does Tomorrow Ever Come? Disaster Narrative and Public Perceptions of Climate Change. *Public Understanding of Science* 15: 435–57.

MacGregor, S. 2010. Gender and Climate Change: From Impacts to Discourses. *Journal of Indian Ocean Region* 6(2): 223–38.

——. 2017. Moving beyond Impacts: More Answers to the "Gender and Climate Change" Question. In S. Buckingham & V. Le Masson (Eds.), *Understanding Climate Change through Gender Relations*, 15–30. London: Routledge.

MacKay, D. 2009. *Sustainable Energy: Without the Hot Air*. Cambridge: UIT Cambridge.

Madii 'Lii Camp. 2016. www.facebook.com/MADII-LII-CAMP-694143797308168/?fref=nf.

Maldonado, J.K., Shearer, C., Bronen, R., Peterson, K. & Lazrus, H. 2013. The Impact of Climate Change on Tribal Communities in the US: Displacement, Relocation, and Human Rights. *Climatic Change* 120(3): 601–14.

Malm, A. 2016a. *Fossil Capital: The Rise of Steam Power and the Roots of Global Warming*. London: Verso.

——. 2016b. Who Lit This Fire? Approaching the History of the Fossil Economy. *Critical Historical Studies* 3: 215–48.

Malm, A. & Hornborg, A. 2014. The Geology of Mankind? A Critique of the Anthropocene Narrative. *The Anthropocene Review* 1: 62–9.

Mamani Ramírez, P. 2011. Qamir qamaña: dureza de "estar estando" y dulzura de "ser siendo." In *Vivir bien: ¿paradigma no-capitalista?* La Paz, Bolivia: CIDES-UMSA.

Manning, M., Lawrence, J., King, D. & Chapman, N. 2015. Dealing with Changing Risks: A New Zealand Perspective on Climate Change Adaptation. *Regional Environmental Change* 15(4): 581–94.

Maraud, S. & Guyot, S. 2016. Mobilization of Imaginaries to Build Nordic Indigenous Natures. *Polar Geography* 39: 196–216.

Marcha Mundial das Mulheres. 2012. *O mundo não é uma mercadoria, as mulheres também não*. Brazil.

Marino, E. & Ribot, J. 2012. Special Issue Introduction: Adding Insult to Injury: Climate Change and the Inequities of Climate Intervention. *Global Environmental Change* 22(2): 323–8.

Matthews, D. 2016. Quantifying Historical Carbon and Climate Debts among Nations. *Nature Climate Change* 6: 60–4.

Mawdsley, E., Savage, L. & Kim, S.M. 2014. A "Post-Aid World"? Paradigm Shift in Foreign Aid and Development Cooperation at the 2011 Busan High Level Forum. *The Geographical Journal* 180(1): 27–38.

McAdam, J. 2012. *Climate Change, Forced Migration, and International Law*. Oxford; New York: Oxford University Press.

McCright, A. 2010. The Effects of Gender on Climate Change Knowledge and Concern in the American Public. *Population and Environment* 32(1): 66–87.

McCright, A. & Dunlap, R. 2011. Cool Dudes: The Denial of Climate Change among Conservative White

Males in the United States. *Global Environmental Change* 21(4): 1163–72.

——. 2015. Bringing Ideology In: The Conservative White Male Effect on Worry about Environmental Problems in the USA. *Journal of Risk Research* 16(2): 211–26.

McKibben, B. 2016a. Global Warming's Terrifying New Chemistry. *The Nation.* March 23, 2016. www.thenation.com/article/global-warming-terrifying-new-chemistry/.

——. 2016b. Why We Need a Carbon Tax, and Why It Won't Be Enough. *Yale Environment 360.* September 12, 2016. http://e360.yale.edu/feature/why_we_need_a_carbon_tax_and_why_it_won_be_enough/3033//.

——. 2017. Stop Swooning over Justin Trudeau: The Man Is a Disaster for the Planet. *The Guardian.* April 17, 2017. www.theguardian.com/commentisfree/2017/apr/17/stop-swooning-justin-trudeau-man-disaster-planet.

McLaren, D., Parkhill, K., Corner, A., Vaughan, N.E. & Pidgeon, N. 2016. Public Conceptions of Justice in Climate Engineering: Evidence from Secondary Analysis of Public Deliberation. *Global Environmental Change* 41: 64–73.

McPherson, G. 2017. Nature Bats Last (blog). https://guymcpherson.com/.

McSheffrey, E. 2016. Trudeau Says Pipelines Will Pay for Canada's Transition to a Green Economy. *National Observer.* March 2, 2016. www.nationalobserver.com/2016/03/02/news/trudeau-says-pipelines-will-pay-canadas-transition-green-economy.

Merchant, C. 1980. *The Death of Nature: Women, Ecology and the Scientific Revolution.* New York: Harper & Row.

Methmann, C. 2010. "Climate Protection" as Empty Signifier: A Discourse Theoretical Perspective on Climate Mainstreaming in World Politics.

Millennium: Journal of International Studies 39(2): 345–72.

Michael, K., Deshpande, T. & Ziervogel, G. 2017. Examining Vulnerability in a Dynamic Urban Setting: The Case of Bangalore's Interstate Migrant Waste Pickers. Social Science Research Network, SSRN. March 2, 2017. https://ssrn.com/abstract=2924375.

Michael, K. & Sreeraj, A.P. 2015. Class Inequality and Climate Change Resilience: Exploring the Nexus in Liberalised India. Social Science Research Network, SSRN. March 5, 2015. https://papers.ssrn.com/sol3/papers.cfm?abstract_id=2572619.

Mieczkowski, Y. 2005. *Gerald Ford and the Challenges of the 1970s.* Lexington, KY: University of Kentucky Press.

Mikaelsson, S. 2014. Winds of Change: The Role and Potential of Sámi Parliamentarians. In J. Gärdebo, M.-B. Öhman & H. Maruyama (Eds.), *Re:Mindings: Con-constituting Indigenous/Academic/Artistic Knowledges,* 79–87. Uppsala: Hugo Valentine Centre, Uppsala University.

Milkoreit, M., Martinez, M. & Eschrich, J. (Eds.). 2016. *Everything Change: An Anthology of Climate Fiction.* Tempe, AZ: Arizona State University. www.dropbox.com/s/gvjk7yvmmbnwkmq/Everything%20Change%20An%20Anthology%20of%20Climate%20Fiction.pdf?dl=0%22.

Miller, D. 1992. Deliberative Democracy and Social Choice. *Political Studies* 40: 54–67.

Mimura, N., Nurse, L.A., McLean, R.F., Agard, J., Briguglio, L., Lefale, P., Payet, R. & Sem, G. (2007). Small Islands. In M.L. Parry, O.F. Canziani, J.P. Palutikof, P.J. van der Linden & C.E. Hanson (Eds.), *Climate Change 2007: Impacts, Adaptation and Vulnerability. Contribution of Working Group II to the Fourth Assessment Report of the Intergovernmental Panel on Climate*

Change, 687–716. Cambridge: Cambridge University Press.

Ministry for the Environment (MfE). 2014. New Zealand Framework for Adapting to Climate Change. INFO 723. Ministry for the Environment, Wellington, NZ. www.mfe.govt.nz/sites/default/files/media/Climate%20Change/nz-framework-for-adapting-to-climate-change-pdf.pdf.

Ministry of Environment and Energy. 2014. Contract Signed for the Construction of Coastal Protection Works in Gaafu Dhaalu Atoll Thinadhoo Island. July 16, 2014. www.environment.gov.mv/v1/news/contract-signed-for-the-construction-of-coastal-protection-works-in-gaafu-dhaalu-atoll-thinadhoo-island/.

——. 2015. Maldives Assumes Chair of Alliance of Small Island States (AOSIS). www.environment.gov.mv/v1/news/maldives-assumes-the-chair-of-alliance-of-small-island-states-aosis/.

Ministry of Environment Energy and Water. 2007. *National Adaptation Programme of Action: Republic of Maldives.* Republic of Maldives: Ministry of Environment Energy and Water. http://unfccc.int/resource/docs/napa/mdv01.pdf.

Ministry of Finance and Treasury. 2017. *State Budget 2017.* Republic of Maldives: Ministry of Finance and Treasury.

Moen, J. 2008. Climate Change: Effects on the Ecological Basis for Reindeer Husbandry in Sweden. *Ambio* 37: 304–11.

Mohamed, H., Naish, A. & Rasheed, Z. 2015. Constitutional Amendment on Foreign Land Ownership up for Debate Tonight. *Minivan News.* July 21, 2015. https://minivannewsarchive.com/politics/constitutional-amendment-on-foreign-land-ownership-up-for-debate-tonight-101163#sthash.xiXhvi3s.3IAfKipU.dpbs.

Monbiot, G. 2016. Neoliberalism: The Ideology at the Root of All Our Problems. *The Guardian.* April 15, 2016. www.theguardian.com/books/2016/apr/15/neoliberalism-ideology-problem-george-monbiot.

——. 2017. Why Are the Crucial Questions about Hurricane Harvey Not Being Asked? *The Guardian.* August 29, 2017. www.theguardian.com/commentisfree/2017/aug/29/hurricane-harvey-manmade-climate-disaster-world-catastrophe.

Moore, J. 2016. Name the System! Anthropocenes and the Capitalocene Alternative. Jason W. Moore. October 9, 2016. https://jasonwmoore.wordpress.com/2016/10/09/name-the-system-anthropocenes-the-capitalocene-alternative/.

Morgan, P. 2006. The Concept of Capacity. European Centre for Development Policy Management. May 2006. http://ecdpm.org/wp-content/uploads/2006-The-Concept-of-Capacity.pdf.

Mörkenstam, U. 2002a. Bilden av den Andre i Svenska Samepolitik. *Svenska Värderingar: Svenska Röda Korset 100 år*, 49–71. Stockholm: Carlssons Förlag.

——. 2002b. Power to Define: The Saami in Swedish Legislation. In K. Karppi & J. Eriksson (Eds.), *Conflict and Cooperation in the North*, 113–45. Umeå: Kulturgräns norr.

Morrison, S.L. 2016. Te Tai Uka a Pia: Iwi Relationships with Antarctic and the Southern Oceans to Enhance Adaptation to Climate Change. Project funded by a Deep South National Science Challenge research grant, Ministry of Business, Innovation, and Employment, Government of New Zealand.

Morrison, S.L. & Vaioleti, T.M. 2012. He Manawa-ā-whenua, e Kore e Mimiti e. Mātauranga Māori and the Springs and Groundwater of the Motueka Area. Commissioned report prepared for Tiakina Te Taiao, Nelson, New Zealand.

Morrow, K. 2017. Changing the Climate of Participation: The Gender Constituency in the Global Climate Change Regime. In S. MacGregor (Ed.), *The Routledge Handbook of Gender and Environment*, 398–411. London: Routledge.

Morton, T. 2013. *Hyperobjects: Philosophy and Ecology after the End of the World*. Minneapolis, MN: University of Minnesota Press.

Mossett, K. 2015. Kandi Mossett Speaks on Suez at #COP21. December 4, 2015. www.youtube.com/watch?v=RAHZ4E7P368.

Mosuela, C. & Matias, D.M. 2014. The Role of a priori Cross-border Migration after Extreme Climate Events: The Case of the Philippines after Typhoon Haiyan. COMCAD Arbeitspapiere Working Papers 126. Universität Bielefeld, Bielefeld, Germany. www.uni-bielefeld.de/tdrc/ag_comcad/downloads/WP_126.pdf.

Munshi, D., Kurian, P., Morrison, T. & Morrison, S.L. 2016. Redesigning the Architecture of Policy-making: Engaging with Maori on Nanotechnology in New Zealand. *Public Understanding of Science* 25(3): 287–302.

Nafiz, A. 2017a. Emboodhoo Lagoon's First Phase: A Three-Island Hard Rock Resort, To Open Next Year. *Maldives Insider*. April 6, 2017. http://maldives.net.mv/17825/emboodhoo-lagoons-first-phase-a-three-island-hard-rock-resort-to-open-next-year/.

——. 2017b. Singapore's Pontiac Kicks Off Major Integrated Tourism Project in the Maldives Lagoon. *Maldives Insider*. May 23, 2017. http://maldives.net.mv/18940/singapores-pontiac-kicks-off-major-integrated-tourism-project-in-maldives-lagoon/.

Naish, A. 2017. Malaysian Company Awarded US$57.5m Project to Build Five Airports. *Maldives Independent*. March 7, 2017. http://maldivesindependent.com/business/malaysian-company-awarded-us57-5m-project-to-build-five-airports-129305.

Nakhooda, S. 2015. Capacity Building Activities in Developing Countries. Presentation at UNFCCC Workshop on Potential Ways to Enhance Capacity-Building Activities. October 17, 2015. http://unfccc.int/files/cooperation_and_support/capacity_building/application/pdf/capacity_building_activities_in_developing_countries.pdf.

Narayan, U. & Harding, S. (Eds.). 2000. *Decentering the Center: Philosophy for a Multicultural, Postcolonial and Feminist World*. Indianapolis, IN: Indiana University Press.

Nasih, A.M. 2017. Parliament Committee Passes Amendment to Slash Petrol-Diesel Duty. *Mihaaru*. June 13, 2017. http://en.mihaaru.com/parliament-committee-passes-amendment-to-slash-petrol-diesel-duty/.

National Institute for Water and Atmospheric Research (NIWA). 2011. Coastal Adaptation to Climate Change: Engaging Communities: Making It Work. CACC125. National Institute for Water and Atmospheric Research, Christchurch, NZ. www.niwa.co.nz/sites/niwa.co.nz/files/making_it_work_final_december2011.pdf.

National Resources Defense Council (NRDC). 2014. *Tar Sands Crude Oil: Health Effects of a Dirty and Destructive Fuel*. February 2014. www.nrdc.org/sites/default/files/tar-sands-health-effects-IB.pdf.

Neimanis, A. & Loewen Walker, R. 2014. Weathering: Climate Change and the "Thick Time" of Transcorporeality. *Hypatia* 29(3): 558–75.

Nel, M. 2015. Coasatu Breaks Stereotypes about Climate Justice. *Mail and Guardian*. March 20, 2015. https://mg.co.za/

article/2015-03-20-00-cosatu-breaks-stereotypes-about-climate-justice.

Newell, P. 2005. Race, Class and the Global Politics of Environmental Inequality. *Global Environmental Politics* 5: 70–94.

Nibert, D. 2002. *Animal Rights / Human Rights: Entanglements of Oppression and Liberation.* New York: Rowman & Littlefield.

Nixon, R. 2013. *Slow Violence and the Environmentalism of the Poor.* Cambridge, MA: Harvard University Press.

Nussbaum, M.C. 2009. *Creating Capabilities: The Human Development Approach.* Cambridge, MA: Harvard University Press.

O'Brien, K. & Wolf, J. 2010. A Values-based Approach to Vulnerability and Adaptation to Climate Change. *Wiley Interdisciplinary Reviews: Climate Change* 1(2): 232–42.

Oka, C.D. 2016. Mothering as Revolutionary Praxis. In A.P. Gumbs, C. Martens & M. Williams (Eds.), *Revolutionary Mothering: Love on the Front Lines*, 51–7. Oakland, CA: PM Press.

Oreskes, N. 2007. The Scientific Consensus on Climate Change: How Do We Know We Are Not Wrong? In J.F.C. DiMento & P. Doughman (Eds.), *Climate Change: What It Means for Us, Our Children, and Our Grandchildren.* Cambridge, MA: MIT Press.

Oreskes, N. & Conway, E. 2011. *The Merchants of Doubt.* New York: Bloomsbury.

Our Children's Trust. 2016. Landmark U.S. *Federal Climate Lawsuit.* www.ourchildrenstrust.org/us/federal-lawsuit/.

Owen, D. 2010. *Green Metropolis: Why Living Smaller, Living Closer, and Driving Less Are the Keys to Sustainability.* New York: Riverhead.

Oxfam. 2014. *Working for the Few: Political Capture and Economic Inequality.* Oxford: Oxfam. www.oxfam.org/en/research/working-few.

——. 2015. Extreme Carbon Inequality. *Oxfam Media Briefing.* December 2, 2015. www.oxfam.org/sites/www.oxfam.org/files/file_attachments/mb-extreme-carbon-inequality-021215-en.pdf.

Pacific Islands Forum. 2013a. *Majuro Declaration for Climate Leadership.* September 5, 2013. http://reliefweb.int/sites/reliefweb.int/files/resources/130905_RMI_PIF_Majuro_Declaration__Commitments.pdf.

——. 2013b. Report on the Majuro Declaration for Climate Leadership. September 5, 2013. http://reliefweb.int/report/world/majuro-declaration-climate-leadership.

Painter, J. 2011. *Poles Apart: The International Reporting of Climate Scepticism.* Oxford: Reuters Institute for the Study of Journalism.

Panko, B. 2016. Humans Have Bogged Down the Earth with 30 Trillion Metric Tons of Stuff, Study Finds. Smithsonian.com. December 9, 2016. www.smithsonianmag.com/science-nature/humans-have-bogged-down-earth-30-trillion-metric-tons-stuff-new-report-finds-180961365/.

Pape, R. & Löffler, J. 2012. Climate Change, Land Use Conflicts, Predation and Ecological Degradation as Challenges for Reindeer Husbandry in Northern Europe: What Do We Really Know after Half a Century of Research? *Ambio* 41: 421–34.

Parenti, C. 2011. *Tropic of Chaos: Climate Change and the New Geography of Violence.* New York: Nation Books.

Parnell, S. & Pieterse, E. 2010. The "Right to the City": Institutional Imperatives of a Developmental State. *International Journal of Urban and Regional Research.* 34(1): 146–62.

Paschen, J.A. & Ison, R. 2014. Narrative Research in Climate Change Adaptation: Exploring a Complementary Paradigm for Research and Governance. *Research Policy* 43(6): 1083–92.

Payne, G. 2016. Fiji Impacted by Tropical Cyclone Zena Just Weeks after Devastation of Tropical Cyclone Winston. News.com.au. April 6, 2016. www.news.com.au/world/fiji-impacted-by-tropical-cyclone-zena-just-weeks-after-devastation-of-tropical-cyclone-winston/news-story/821fea3 87ebf90cd6ebb885f23114b1c.

Payne, V. 2016. Co-governance Delivers Benefits for Waikato Rivers. *Waikato Times*. September 14, 2016. www.stuff.co.nz/waikato-times/opinion/84029610/Opinion-Co-governance-delivers-benefits-for-Waikato-rivers.

Pearse, R. 2017. Gender and Climate Change. *WIREs Climate Change* 8(2): e451.

Pease, B. 2016. Masculinism, Climate Change and "Man-made" Disasters: Towards an Environmental Profeminist Response. In E. Enarson & B. Pease (Eds.), *Men, Masculinities and Disaster*, 21–33. London: Routledge.

Peden, M., Scurfield, R., Sleet, D., Mohan, D., Hyder, A., Jarawan, E. & Mathers, C. (Eds.). 2004. *World Report on Road Traffic Injury Prevention*. World Health Organization. http://apps.who.int/iris/bitstream/10665/42871/1/9241562609.pdf.

Pellow, D.N. 2014. *Total Liberation: The Power and Promise of Animal Rights and the Radical Earth Movement*. Minneapolis, MN: University of Minnesota Press.

——. 2016. Toward a Critical Environmental Justice Studies: Black Lives Matter as an Environmental Justice Challenge. *DuBois Review* 13(2): 221–36.

Peoples' Agreement. 2010. World People's Conference on Climate Change and the Rights of Mother Earth. https://pwccc.wordpress.com/support/.

Persson, M. & Öhman, M.-B. 2014. Visions for the Future at the Source of the Ume River, Sweden: The Battle against the Rönnbäck Nickel Mining Project. In J. Gärdebo, M.-B. Öhman & H. Maruyama (Eds.), *Re:Mindings: Co-constituting Indigenous/Academic/Artistic Knowledges*, 103–19. Uppsala: Hugo Valentine Centre, Uppsala University.

Peterson, V.S. & Runyan, A.S. 2013. *Global Gender Issues in the New Millennium*, 4th edition. Boulder, CO: Westview Press.

Petheram, Z., Campbell, B., High, C. & Stacey, N. 2010. "Strange Changes": Indigenous Perspectives of Climate Change and Adaptation in NE Arnhem Land (Australia). *Global Environmental Change* 20(4): 681–92.

Petroff, A. 2017. CEOs Make a Final Urgent Plea: Don't Pull Out of Paris Accord. *CNN*. June 1, 2017. http://money.cnn.com/2017/06/01/news/trump-paris-climate-deal-business-ceo/index.html?iid=EL.

Phakathi, M. 2017. Universities in Global South Aim to End Reliance on Western Experts. *Climate Home*. July 4, 2017. www.climatechangenews.com/2017/07/04/universities-global-south-aim-end-reliance-western-experts/.

Pharr, S. 1988. *Homophobia: A Weapon of Sexism*. Little Rock, AR: The Women's Project.

Picketty, T. 2014. *Capital in the Twenty-First Century*. Translated by A. Goldhammer. Cambridge, MA: Belknap Press of Harvard University Press.

Pidcock, R. 2016. Analysis: What Global Emissions in 2016 Mean for Climate Change Goals. *Carbon Brief*. November 15, 2016. www.carbonbrief.org/what-global-co2-emissions-2016-mean-climate-change?utm_source=Weekly+Carbon+Briefing&

utm_campaign=2a886c0482-Carbon_Brief_Weekly_18_11_2016&utm_medium=email&utm_term=0_3ff5ea836a-2a886c0482-303452073&ct=t.

Plumwood, V. 1993. *Feminism and the Mastery of Nature*. London: Routledge.

Poell, T. & van Dijck, J. 2015. Social Media and Activist Communication. In C. Atton (Ed.), *The Routledge Companion to Alternative and Community Media*, 527–37. New York: Routledge.

Porritt, J. 2013. *The World We Made: Alex McKay's Story from 2050*. New York: Phaedon Press.

Poushter, J. 2016. Americans, Canadians Differ in Concern about Climate Change. *Pew Research Center*. March 9, 2016. www.pewresearch. org/fact-tank/2016/03/09/americans-canadians-differ-in-concern-about-climate-change/.

Prasad, V. 2014. *The Poorer Nations: A Possible History of the Global South*. London: Verso.

Principles of Environmental Justice. 1991. EJNet.org. April 6, 1996. www.ejnet. org/ej/principles.pdf.

Prison Ecology Project. 2016. Examples of Prison Pollution and Environmental Justice Issues in U.S. Prisons. PEP, Lake Worth, Florida. http://nationinside. org/campaign/prison-ecology/.

Project Drawdown. 2017. www. drawdown.org/.

Public Policy Institute of California (PPIC). 2016. California's Latest Drought. July 2016. www.ppic.org/main/publication_show.asp?i=1087.

Puig de la Bellacasa, M. 2012. "Nothing Comes without Its World": Thinking with Care. *The Sociological Review* 60(2): 197–215.

Pynn, L. 2012. Metro Vancouver Dike Improvements Could Cost $9.5 Billion by 2100: New Report. *Vancouver Sun*. December 12, 2012. www. vancouversun.com/technology/

Metro+Vancouver+dike+improvements+could+cost+billion+2100+report+with+video/7682197/story.html.

Rakova, U. 2012. The Carterets Islands Integrated Relocation Program, Tinputz, Bougainville, Papua New Guinea. Presented at the Catholic Diocese of Australia Climate Change Meeting, Brisbane, Australia.

Ramesh, R. 2009. Maldives Ministers Prepare for Underwater Cabinet Meeting. *The Guardian*. October 7, 2009. www.theguardian.com/world/2009/oct/07/maldives-underwater-cabinet-meeting.

Ransom, J.W. 1999. The Waters. In *Haudenosaunee Environmental Task Force: Words That Come before All Else: Environmental Philosophies of the Haudenosaunee*, 25–43. Cornwall Island, ON: Native North American Travelling College.

Raskin, P. 2016. Journey to Earthland: The Great Transition to Planetary Civilization. Tellus Institute. www. greattransition.org/documents/Journey-to-Earthland.pdf.

Raworth, K. 2014. Must the Anthropocene Be a Manthropocene? *The Guardian*. October 20, 2014. www.theguardian. com/commentisfree/2014/oct/20/anthropocene-working-group-science-gender-bias.

Reeves, R. 2017. Stop Pretending You're Not Rich. *The New York Times*. June 10, 2017. www.nytimes. com/2017/06/10/opinion/sunday/stop-pretending-youre-not-rich. html?_r=0.

Resurrección, B. 2017. Gender and Environment in the Global South: From "Women, Environment and Development" to Feminist Political Ecology. In S. MacGregor (Ed.), *The Routledge Handbook of Gender and Environment*, 71–85. London: Routledge.

Revi, A. 2008. Climate Change Risk: An Adaptation and Mitigation Agenda for Indian Cities. *Environment and Urbanization* 20(1): 207–29.

Ribot, J. 2010. Vulnerability Does Not Fall from the Sky: Toward Multiscale, Pro-poor Climate Policy. *Social Dimensions of Climate Change: Equity and Vulnerability in a Warming World* 2: 47–74.

——. 2014. Cause and Response: Vulnerability and Climate in the Anthropocene. *The Journal of Peasant Studies* 41(5): 667–705.

Richards, G. 2010. Queering Katrina: Gay Discourses of the Disaster in New Orleans. *Journal of American Studies* 44(3): 519–34.

Rigg, K. 2013. Island Nations Lead Where All Must Follow in a Looming Climate Crisis. *HuffPost.* September 2, 2013. www.huffingtonpost.com/kelly-rigg/island-nations-lead-where_b_3854054.html.

Rising Tide North America. 2015. Climate Justice. https://risingtidenorthamerica.org/about-rising-tide-north-america/climate-justice/.

Rivera, S. 1991. Pachakuti: los horizontes históricos del colonialismo interno. *NACLA como "Aymara Past, Aymara Future"* 25(3): 18–45.

Rivera Cusicanqui, S. 2010. *Violencias (re) encubiertas en Bolivia.* La Paz: Mirada Salvaje.

RNZ. 2013. Pacific Leaders Put Out Climate Change Declaration as Summit Ends. September 5, 2013. www.radionz.co.nz/international/pacific-news/220672/pacific-leaders-put-out-climate-change-declaration-as-summit-ends.

Robb, P. 2015. Q and A: Sheila Watt-Cloutier Seeks Some Cold Comfort. *Ottawa Citizen.* March 27, 2015. http://ottawacitizen.com/entertainment/books/q-and-a-sheila-watt-cloutier-seeks-some-cold-comfort.

Roberts, J.T. & Parks, B. 2006. *A Climate of Injustice: Global Inequality, North–South Politics, and Climate Policy.* Cambridge, MA: MIT Press.

Rockefeller Foundation. 2017. 100 Resilient Cities. www.100resilientcities.org.

Rogers, P. 2016. Fracking Ban: Environmentalists Declare Victory on Monterey Measure Z. *San Jose Mercury News.* November 9, 2016. www.mercurynews.com/2016/11/09/fracking-ban-environmentalists-declare-victory-on-monterey-measure-z/.

Rogers, S. & Xue, T. 2015. Resettlement and Climate Change Vulnerability: Evidence from Rural China. *Global Environmental Change: Human and Policy Dimensions* 35: 62–9.

Roosvall, A. & Tegelberg, M. 2013. Framing Climate Change and Indigenous Peoples: Intermediaries of Urgency, Spirituality and De-nationalization. *The International Communication Gazette* 75: 392–409.

Rose, C. 2017. Breitbart's Bannon Declares War on the GOP. *60 Minutes.* September 20, 2017. www.cbsnews.com/news/60-minutes-breitbart-steve-bannon-declares-war-on-the-gop/.

Roughgarden, J. 2013. *Evolution's Rainbow: Diversity, Gender, and Sexuality in Nature and People,* 2nd edition. Berkeley, CA: University of California Press.

Roy, M., Hulme, D. & Jahan, F. 2013. Contrasting Adaptation Responses by Squatters and Low Income Tenants in Khulna, Bangladesh. *Environment and Urbanization* 25(10): 157–67.

Royster, J. 1997. Oil and Water in Indian Country. *Natural Resources Journal* 37: 457–90. http://digitalrepository.unm.edu/cgi/viewcontent.cgi?article=1694&context=nrj.

Russell, S., Greenaway, A., Carswell, F. & Weaver, S. 2014. Moving beyond "Mitigation and Adaptation": Examining Climate Change Responses in New Zealand. *Local Environment* 19(7): 767–85.

Ryan, T. 1984. *Palagi Views of Niue: Historical Literature 1774–1899*. Compiled by T.F. Ryan. Auckland: Auckland University Bindery.

Sakakibara, C. 2017. People of the Whales: Climate Change and Cultural Resilience among Iñupiat of Arctic Alaska. *Geographical Review* 107(1): 159–84.

Saki, O. 2017. How Companies Are Using Lawsuits to Silence Environmental Activists: And How Philanthropy Can Help. The Ford Foundation. June 13, 2017. www.fordfoundation.org/ideas/equals-change-blog/posts/how-companies-are-using-law-suits-to-silence-environmental-activists-and-how-philanthropy-can-help/.

Saleem, A. 2016. *Environmental Impact Assessment for Coastal Protection at Gn. Fuvahmulah*. Malé, Maldives: Maldives Energy and Environment Company.

Sampei, Y. & Aoyagi-Usui, M. 2009. Mass-Media Coverage, Its Influence on Public Awareness of Climate Change Issues, and Implications for Japan's National Campaign to Reduce Greenhouse Gas Emissions. *Global Environmental Change* 19: 203–12.

Saro-Wiwa, K. 1992. *Genocide in Nigeria: The Ogoni Tragedy*. Port Harcourt, Nigeria: Saros International.

Schäfer, M.S. & Schlichting, I. 2014. Media Representations of Climate Change: A Meta-Analysis of the Research Field. *Environmental Communication* 8(2): 142–60.

Scheer, H. 2012. *The Energy Imperative: 100 Per Cent Renewable Now*. London: Earthscan.

Schlosberg, D. 2004. Reconceiving Environmental Justice: Global Movements and Political Theories. *Environmental Politics* 13(3): 517–40.

——. 2012. Climate Justice and Capabilities: A Framework for Adaptation Policy. *Ethics and International Affairs* 26(4): 445–61.

Schlosberg, D. & Collins, L.B. 2014. From Environmental to Climate Justice: Climate Change and the Discourse of Environmental Justice. *WIREs Climate Change* 5: 359–74.

Schmidt, A. & Schäfer, M.S. 2015 Constructions of Climate Justice in German, Indian and US Media. *Climatic Change* 133(3): 535–49.

Science Learning Hub. 2016. New Zealand's National Science Challenges. www.sciencelearn.org.nz/resources/1112-new-zealand-s-national-science-challenges.

Secretariat of the Pacific Environment Programme. 2017. Climate Change: Overview. www.sprep.org/Climate-Change/climate-change-overview.

Sekerci, Y. & Petrovskii, S. 2015. Mathematical Modelling of Plankton–Oxygen Dynamics under the Climate Change. *Bulletin of Mathematical Biology* 77(12): 2325–53.

Selby, R., Mulholland, M. & Moore, P. 2010. Introduction. In S.R. Selby, P. Moore & M. Mulholland (Eds.), *Maori and the Environment: Kaitiaki*. Wellington, New Zealand: Huia Publishers.

Sen, A. 2009. *The Idea of Justice*. Cambridge, MA: Belknap Press of Harvard University Press.

Shaig, A. 2006. *Climate Change Vulnerability and Adaptation Assessment of Maldives Land and Beaches*. Townsville, Australia: James Cook University.

Sharife, K. 2011. Colonising Africa's Atmospheric Commons. In Patrick Bond (Ed.), *Durban's Climate Gamble: Trading Carbon, Betting the Earth*, 157–78. Pretoria: Unisa Press.

Sharpe, C. 2016. *In the Wake: On Blackness and Being*. Durham, NC: Duke University Press.

Shearer, C. 2011. *Kivalina: A Climate Change Story*. Chicago, IL: Haymarket Books.

Shearman, D.J.C. & Smith, J.W. 2007. *The Climate Change Challenge and the*

Failure of Democracy. Westport, CT: Praeger.

Sheppard, S., Shaw, A., Flanders, D., Burch, S., Wiek, A., Carmichael, J., Robinson, J. & Cohen, S. 2011. Future Visioning of Local Climate Change: A Framework for Community Engagement and Planning with Scenarios and Visualisation. *Futures* 43(4): 400–12.

Shi, L., Chu, E., Anguelovski, I., Aylett, A., Debats, J., Goh, K., Schenk, T., Seto, K., Dodman, D., Roberts, D. & Roberts, J.T. 2016. Roadmap towards Justice in Urban Climate Adaptation Research. *Nature Climate Change* 6(2): 131–7.

Shilling, V. 2013. Woman with Eagle Feather: The Photo "Heard Round the World." *Indian Country Today.* November 21, 2013. http://indiancountrytodaymedianetwork.com/2013/11/21/woman-eagle-feather-photo-heard-round-world-152357.

Shingler, B. & Smith, S. 2016. NEB Cancels 2 Days of Energy East Hearings in Montreal after "Violent Disruption." *CBC News.* August 29, 2016. www.cbc.ca/news/canada/montreal/neb-hearings-energy-east-protest-quebec-2016-1.3739215.

Shinglespit Consultants Incorporated. 2017. What Is the Social License? https://socialicense.com/definition.html.

Singer, P. 2004. *One World: The Ethics of Globalization*, 2nd edition. New Haven, CT: Yale University Press.

Singh, C., Basu, R. & Srinivas, A. 2016. Livelihood Vulnerability and Adaptation in Kolar District, Karnataka, India: Mapping Risks and Responses. ASSAR Short Report. University of Cape Town, Cape Town, South Africa.

Skarin, A., Nelleman, C., Sandström, P., Rönnegård, L. & Lundqvist, H. 2013. Renar och Vindkraft: Studie Från Anläggningen av två Vindkraftparker i Malå Sameby. Vindval Rapport 6564. Stockholm: Naturvårdsverket. www.naturvardsverket.se/Documents/publikationer6400/978-91-620-6564-5.pdf.

Small, G. 1994. War Stories: Environmental Justice in Indian Country. *Daybreak* 4(2): no pagination.

Smith, Deirdre. 2014. Why the Climate Movement Must Stand with Ferguson. 350.org. August 20, 2014. http://350.org/how-racial-justice-is-integral-to-confronting-climate-crisis/.

Smith, Dinitia. 2004. Love That Dare Not Squeak Its Name. *The New York Times.* February 7, 2004. www.nytimes.com/2004/02/07/arts/love-that-dare-not-squeak-its-name.html?_r=0.

Smith, H. 2014. Is Necessity the Mother of Climate Protest? *Grist.* September 8, 2014. http://grist.org/climate-energy/is-necessity-the-mother-of-climate-protest/?utm_source=newsletter&utm_medium=email&utm_term=Daily%2520Sept%25209&utm_campaign=daily.

Smith, R. 2014. Green Capitalism: The God That Failed. *Truthout.* January 9, 2014. www.truth-out.org/news/item/21060-green-capitalism-the-god-that-failed.

Smoke and Fumes. 2016. Fumes. www.smokeandfumes.org/fumes.

Solchaga, F. 2016. Spain. In G. Picton-Turbervill & J. Derrick (Eds.), *Energy 2017.* London: Global Legal Insights. www.globallegalinsights.com/practice-areas/energy/global-legal-insights---energy-5th-ed./spain#chaptercontent1.

Solnit, R. 2009. *A Paradise Built in Hell: The Extraordinary Communities That Arise in Disaster.* New York: Viking.

——. 2012a. Men Explain Things to Me. *Guernica: A Magazine of Global Arts and Politics.* August 20, 2012. www.guernicamag.com/daily/rebecca-solnit-men-explain-things-to-me/.

——. 2012b. The Sky's the Limit: The Demanding Gifts of 2012. *TomDispatch.* December 23, 2012.

www.tomdispatch.com/blog/175632/
tomgram%3A_rebecca_solnit_2013_as_
year_zero_for_us_–_and_our_planet.

——. 2015. The Most Important Thing
We Can Do to Fight Climate Change
Is Try. *The Nation.* March 23, 2015.
www.thenation.com/article/198537/
unpredictable-weather#.

Solomon, S., Qin, D., Manning, M.,
Chen, Z., Marquis, M., Averyt, K.B.,
Tignor, M. & Miller, H.L. (Eds.). 2007.
*Contribution of Working Group I to
the Fourth Assessment Report of the
Intergovernmental Panel on Climate
Change 2007.* Cambridge: Cambridge
University Press. www.ipcc.ch/
publications_and_data/ar4/wg1/en/
contents.html.

Stalley, P. 2013. Principled Strategy:
The Role of Equity Norms in China's
Climate Change Diplomacy. *Global
Environmental Politics* 13: 1–4.

Steffen, W., Richardson, K., Rockström, J.,
Cornell, S.E., Fetzer, I., Bennett, E.M.,
Biggs, R., Carpenter, S.R., de Vries, W.,
de Wit, C.A., Folke, C., Gerten, D.,
Heinke, J., Mace, G.M., Persson, L.M.,
Ramanathan, V., Reyers, B. & Sörlin, S.
2015. Planetary Boundaries:
Guiding Human Development on a
Changing Planet. *Science.* January
15, 2015. http://science.sciencemag.
org/content/early/2015/01/14/
science.1259855.

Steffen, W., Sanderson, R.A., Tyson, P.D.,
Jäger, J., Matson, P.A., Moore III, B.,
Oldfield, F., Richardson, K.,
Schellnhuber, H.-J., Turner, B.L. &
Wasson, R.J. 2004. *Global Change
and the Earth System: A Planet under
Pressure.* Berlin: Springer-Verlag.

Steinhardt, H.C. & Wu, F.S. 2016. In the
Name of the Public: Environmental
Protest and the Changing Landscape
of Popular Contention in China. *China
Journal* 75: 61–82.

Stephens, B. & Sprinkle, A. 2013. *Goodbye
Gauley Mountain: An Ecosexual Love
Story.* Film. San Francisco: Fecund Arts.

Stephens, B. & Sprinkle, A. 2016.
Ecosexuality. In R.C. Hoogland
(Ed.), *Gender: Nature*, 313–30.
London: Macmillan Interdisciplinary
Handbooks.

Stephenson, W. 2015. *What We Are Fighting
for Now Is Each Other: Dispatches
from the Frontlines of Climate Justice.*
Boston, MA: Beacon Press.

Stern, T. 2017. Trump Is Wrong on
the Paris Climate Agreement. I
Know Because I Negotiated It.
Washington Post. May 31, 2017.
www.washingtonpost.com/
opinions/trump-is-wrong-on-the-
paris-climate-agreement-i-know-
because-i-negotiated-it/2016/05/31/
ce3a680a-2667-11e6-ae4a-
3cdd5fe74204_story.
html?utm_term=.4826b784545d.

Stevenson, H. 2016. The Wisdom of the
Many in Global Governance: An
Epistemic-Democratic Defense of
Diversity and Inclusion. *International
Studies Quarterly* 60(3): 400–12.

Stevenson, H. & Dryzek, J.S. 2014.
*Democratizing Global Climate
Governance.* New York: Cambridge
University Press.

Stiglitz, J. 2008. Making Globalization
Work: The 2006 Geary Lecture. *The
Economic and Social Review* 39(3):
171–90.

——. 2017. Trump's Reneging on Paris
Climate Deal Turns the US into a
Rogue State. *The Guardian.* June
2, 2017. www.theguardian.com/
business/2017/jun/02/paris-climate-
deal-to-trumps-rogue-america.

Storey, B., Noy, I., Townsend, W., Kerr, S.,
Salmon, R., Middleton, D., Filippova, O.
& James, V. 2017. Insurance, Housing
and Climate Adaptation: Current
Knowledge and Future Research.
MOTU Note # 27. http://motu.
nz/assets/Documents/our-work/
environment-and-resources/
climate-change-impacts/Insurance-
Housing-and-Climate-Adaptation2.pdf.

Suagee, Dean. 1994. Turtle's War Party: An Indian Allegory on Environmental Justice. *Journal of Environmental Law and Litigation* 9: 461–98.

Sudbury, J. 2009. Challenging Penal Dependency: Activist Scholars and the Antiprison Movement. In J. Sudbury & M. Okazawa-Rey (Eds.), *Activist Scholarship: Anti-Racism, Feminism, and Social Change*, 17–35. Boulder, CO: Paradigm Publishers.

Sultana, F. 2014. Gendering Climate Change: Geographical Insights. *The Professional Geographer* 66(3): 372–81.

Sutikalh. 2014. Supreme Court of Canada Rules on Landmark Decision on Aboriginal Title. June 26, 2014. http://sutikalh.blogspot.ca/.

Swedish Radio. 2017. Sameby får FN-stöd mot Gruvplanerna i Rönnbäck. May 4, 2017. https://sverigesradio.se/sida/artikel.aspx?programid=2327&artikel=6701302.

Swyngedouw, E. 2010. Apocalypse Forever? Post-political Populism and the Spectre of Climate Change. *Theory, Culture and Society* 27(2–3): 213–32.

Tauli-Corpuz, V. & Lynge, A. 2008. Impact of Climate Change Mitigation Measures on Indigenous Peoples and on Their Territories and Lands. Paper submitted to Permanent Forum on Indigenous Issues, Seventh session. New York: UN Permanent Forum on Indigenous Issues.

The Movement for Black Lives. 2016. A Vision for Black Lives: Policy Demands for Black Power, Freedom, and Justice. www.policy.m4bl.org.

The President's Office. 2017. Statement by H.E Abdulla Yameen Abdul Gayoom, President of Maldives on the United States Withdrawal from Paris Agreement. The Presidency, Republic of Maldives. June 2, 2017. www.presidencymaldives.gov.mv/Index.aspx?lid=11&dcid=17938.

The Star. 2016. Canada Needs an Ambitious National Climate Plan. Editorial. September 19, 2016. www.thestar.com/opinion/editorials/2016/09/19/canada-needs-an-ambitious-national-climate-plan-editorial.html.

Thompson, A. 2017. A Coal Plant in India Is Turning CO2 into Baking Soda. January 5, 2017. www.popularmechanics.com/science/green-tech/a24631/india-co2-baking-soda/.

Tiu, S.A. 2016. Traditional Ecological Knowledge in Sustainable Resource Management in Papua New Guinea: The Role of Education and Implications for Policy. PhD Thesis. Hamilton, NZ: University of Waikato. http://hdl.handle.net/10289/10704.

Torp, E. 2014. Det Rättsliga Skyddet av Samisk Renskötsel. *Svensk Juristtidning*: 122–48.

Treaty Alliance Against Tar Sands Expansion. 2016. www.treatyalliance.org/.

Treaty Indian Tribes in Western Washington. 2011. *Treaty Rights at Risk: Ongoing Habitat Loss, the Decline of the Salmon Resource, and Recommendations for Change*. Olympia, WA: Northwest Indian Fisheries Commission. http://nwifc.org/w/wp-content/uploads/downloads/2011/08/whitepaper628finalpdf.pdf.

Trump, D. 2012. The Concept of Global Warming ... Twitter. November 6, 2012.

——. 2017. Statement by President Trump on the Paris Climate Accord. The White House. June 1, 2017. www.whitehouse.gov/the-press-office/2017/06/01/statement-president-trump-paris-climate-accord.

Tsagas, I. 2015. Spain Approves "Sun Tax," Discriminates against Solar PV. *Renewable Energy World*. October 23, 2015. www.renewableenergyworld.com/articles/2015/10/spain-approves-sun-tax-discriminates-against-solar-pv.html.

Tsolkas, P. 2015. Mass Incarceration vs. Rural Appalachia. *Earth Island Journal.* August 24, 2015. www.earthisland.org/journal/index.php/elist/eListRead/mass_incarceration_vs._rural_appalachia/.

Turkewitz, J. 2017. Tribes That Live Off Coal Hold Tight to Trump's Promises. *The New York Times.* April 1, 2017. www.nytimes.com/2017/04/01/us/trump-coal-promises.html?mcubz=3.

UNFCCC. 2014. Glossary of Climate Change Acronyms and Terms: Capacity Building. United Nations Framework Convention on Climate Change. http://unfccc.int/essential_background/glossary/items/3666.php.

——. 2015. *Paris Agreement.* United Nations Framework Convention on Climate Change. http://unfccc.int/files/essential_background/convention/application/pdf/english_paris_agreement.pdf.

Unist'ot'en Camp. 2016. www.facebook.com/unistoten.

United Nations. 2008. *United Nations Declaration on the Rights of Indigenous Peoples.* www.un.org/esa/socdev/unpfii/documents/DRIPS_en.pdf.

United Nations Development Programme (UNDP). 1999. *Human Development Report 1999.* New York: UNDP.

United Nations WomenWatch. 2009. Women, Gender Equality and Climate Change. www.un.org/womenwatch/feature/climate_change/downloads/Women_and_Climate_Change_Factsheet.pdf.

University of Leicester. 2015. Failing Phytoplankton, Failing Oxygen: Global Warming Disaster Could Suffocate Life on Planet Earth. *Science Daily.* December 1, 2015. www.sciencedaily.com/releases/2015/12/151201094120.htm.

Uniyal, M. 2017. The Narmada Bachao Andolan Legacy and How It Created a Generation of Activists in India. Your Story. June 15, 2017. https://yourstory.com/2017/06/narmada-bachao-andolan-legacy/.

Urry, J. 2011. *Climate Change and Society.* Cambridge: Polity.

US Energy Information Administration. 2016. United States Remains Largest Producer of Petroleum and Natural Gas Hydrocarbons. May 23, 2016. www.eia.gov/todayinenergy/detail.cfm?id=26352.

Vakulabharanam, V. & Motiram, S. 2012. Understanding Poverty and Inequality in Urban India since Reforms: Bringing Quantitative and Qualitative Approaches Together. *Economic and Political Weekly* 47(47–48): 44–52.

Valenza, A. 2010. Bolivian President: Eating Estrogen-Rich Chicken Makes You Gay. *International Lesbian Gay Association.* April 25, 2010. http://ilga.org/bolivian-president-eating-estrogen-rich-chicken-makes-you-gay/.

Valerio, K.A. 2014. Storm of Violence, Surge of Struggle: Women in the Aftermath of Typhoon Haiyan (Yolanda). *Asian Journal of Women's Studies* 20(1): 148–63.

van Dooren, T. 2014. Care. *Environmental Humanities Living Lexicon* 5: 291–4. http://environmentalhumanities.org/arch/vol5/5.18.pdf.

Van Reybrouck, D. 2016. We Have One Year to Make Democracy Work in Europe. Or Else the Trumps Take Over. *The Correspondent.* November 19, 2016. https://thecorrespondent.com/5711/we-have-one-year-to-make-democracy-work-in-europe-or-else-the-trumps-take-over/570854427-59d8ed4b.

van Wyck, P. 2010. *Highway of the Atom.* Montreal; Kingston, ON: McGill-Queens University Press.

VanDeveer, S.D. & Dabelko, G.D. 2001. It's Capacity, Stupid: International Assistance and National

Implementation. *Global Environmental Politics* 1(2): 18–29.

Vaughan, G. 2002. Mothering, Co-muni-cation, and the Gifts of Language. In E. Wyschogrod, J.-J. Goux & E. Boynton (Eds.), *The Enigma of Gift and Sacrifice*, 91–113. New York: Fordham University Press.

Vencatesan, J., Daniels, R.J.R., Jayaseelan, J.S. & Karthick, N.M. 2014. Comprehensive Management Plan for Pallikaranai Marsh 2014–2019. Care Earth Trust.

Vinyeta, K. & Lynn, K. 2013. Exploring the Role of Traditional Ecological Knowledge in Climate Change Initiatives. General Technical Report No. PNW-GTR-879. United States Department of Agriculture, Portland, OR.

Vu, T.B. & Noy, I. 2015. Regional Effects of Natural Disasters in China: Investing in Post-disaster Recovery. *Natural Hazards* 75: S111–S126.

Wadhams, P. 2017. *Farewell to Ice: A Report from the Arctic*. London: Allen Lane.

Wallace-Wells, D. 2017. The Uninhabitable Earth. *New York Magazine*. July 9, 2017. http://nymag.com/daily/intelligencer/2017/07/climate-change-earth-too-hot-for-humans.html.

Wang, C. & Chen, Y. 2013. On Climate Justice. *International Social Science Journal* 2. [Chinese edition.]

——. 2014. Climate Justice and the Target and System Selection of China's Legislation on Addressing Climate Change. *China Higher Education Social Science* 2014(2): 125–60. [In Chinese.]

Wang, J., Brown, D.G. & Agrawal, A. 2013. Climate Adaptation, Local Institutions, and Rural Livelihoods: A Comparative Study of Herder Communities in Mongolia and Inner Mongolia, China. *Global Environmental Change: Human and Policy Dimensions* 23: 1673–83.

Wapner, P. 2014. The Changing Nature of Nature: Environmental Politics in the Anthropocene. *Global Environmental Politics* 14(4): 36–54.

Warlenius, R. 2016. Linking Ecological Debt and Ecologically Unequal Exchange: Stocks, Flows, and Unequal Sink Appropriation. *Journal of Political Ecology* 23(2016): 364–80.

——. 2017. Asymmetries: Conceptualizing Environmental Inequalities as Ecological Debt and Ecologically Unequal Exchange. PhD Thesis. Lund: Lund University.

Warlenius, R., Pierce, G. & Ramasar, V. 2015. Reversing the Arrow of Arrears: The Concept of "Ecological Debt" and Its Value for Environmental Justice. *Global Environmental Change* 30: 21–30.

Wates, J. 2017. EEB Calls for Tough Response to US Withdrawal from Paris Agreement. EEB European Environmental Bureau. June 2, 2017. eeb.org/eeb-calls-for-tough-response-to-us-withdrawal-from-paris-agreement/.

Wearden, G. 2014. Oxfam: 85 Richest People as Wealthy as Poorest Half of the World. *The Guardian*. January 20, 2014. www.theguardian.com/business/2014/jan/20/oxfam-85-richest-people-half-of-the-world.

Weaver, J. 1996. *Defending Mother Earth: Native American Perspectives on Environmental Justice*. Maryknoll, NY: Orbis Books.

Webb, A.P. & Kench, P.S. 2010. The Dynamic Response of Reef Islands to Sea-Level Rise: Evidence from Multi-decadal Analysis of Island Change in the Central Pacific. *Global and Planetary Change* 72(3): 234–46.

Wei, Y.M., Wang, K., Wang, Z.H. & Tatano, H. 2015. Vulnerability of Infrastructure to Natural Hazards and Climate Change in China. *Natural Hazards* 75: S107–S110.

Weissman, R. 2017. CorporateCabinet.org Tracks Trump Cabinet Picks: New Website Highlights Corporate

Ties, Potential Conflicts of Interest of Top Political Appointees. Public Citizen. January 6, 2017. www.citizen.org/media/press-releases/corporatecabinetorg-tracks-trump-cabinet-picks.

West, T. 2012. Environmental Justice and International Climate Change Legislation: A Cosmopolitan Perspective. *Georgetown International Environmental Law Review* 25(1): 129–74.

Whyte, K.P. 2017. Indigenous Climate Justice Teaching Materials and Advanced Bibliography. http://kylewhyte.cal.msu.edu/climate-justice/.

——. 2019. Indigeneity in Geoengineering Discourses: Some Considerations. *Ethics, Policy & Environment*: 1–19.

Wiedenhofer, D., Guan, D., Liu, Z., Meng, J., Zhang, N. & Wei, Y.-M. 2017. Unequal Household Carbon Footprints in China. *Nature Climate Change* 7(1): 75–80.

Wike, R., Stokes, B., Poushter, J. & Fetterolf, J. 2017. U.S. Image Suffers as Publics around World Question Trump's Leadership. Pew Research Center. June 26, 2017. www.pewglobal.org/2017/06/26/u-s-image-suffers-as-publics-around-world-question-trumps-leadership/.

Wikipedia. 2015. Wicked Problem. https://en.wikipedia.org/wiki/Wicked_problem.

Wildcat, D.R. 2009. *Red Alert! Saving the Planet with Indigenous Knowledge*. Golden, CO: Fulcrum.

Williams, A. 2015. Afrikan Black Coalition Accomplishes UC Prison Divestment! Afrikan Black Coalition. December 12, 2015. http://afrikanblackcoalition.org/2015/12/18/afrikan-black-coalition-accomplishes-uc-prison-divestment/.

Williams, G.A., Helmuth, B., Russell, B.D., Dong, Y., Thiyagarajan, V. & Seuront, L. 2016. Meeting the Climate Change Challenge: Pressing Issues in Southern China and Southeast Asian Coastal Ecosystems. *Regional Studies in Marine Science*. http://dx.doi.org/10.1016/j.rsma.2016.07.002.

Williams, R. 1960. *Culture and Society 1780–1950*. London: Chatto & Windus.

Williams, T. & Hardison, P. 2013. Culture, Law, Risk and Governance: Contexts of Traditional Knowledge in Climate Change Adaptation. *Climatic Change* 120(3): 531–44.

Wisner, B. 2010. Climate Change and Cultural Diversity. *International Social Science Journal* 61(199): 131–40.

Wolf, M. 2017. Donald Trump's Bad Judgment on the Paris Accord. *Financial Times*. June 6, 2017. www.ft.com/content/eecc80f6-4936-11e7-a3f4-c742b9791d43.

World Bank. 2015. Project Appraisal Document Maldives: Climate Change Adaptation Project (CCAP). World Bank Group, Washington, DC. http://documents.worldbank.org/curated/en/468281467986350615/pdf/PAD1258-PAD-P153301-Box393233B-PUBLIC-CCAP-PAD-15-May-final.pdf.

——. 2017. The World Bank in Pacific Islands: Overview. April 10, 2017. www.worldbank.org/en/country/pacificislands/overview.

World Health Organization (WHO). 2017. Road Traffic Injuries. www.who.int/mediacentre/factsheets/fs358/en/.

Wu, F. 2013. Environmental Activism in Provincial China. *Journal of Environmental Policy and Planning* 15: 89–108.

Xi, J. 2015. Build Up a Win-Win, Fair and Reasonable Collaborative Mechanism to Govern Climate Change (xieshou goujian hezuo gongying, gongping heli de qihou bianhua zhili jizhi). Speech given at the COP 21 Conference in Paris, November 30, 2015.

York, C. 2015. Muslims Helping with UK Flood Response but Britain First and EDL Nowhere to Be Seen. *HuffPost United Kingdom*.

December 28, 2015. www. huffingtonpost.co.uk/2015/12/28/ muslims-flooding-britain-first-english-defence-league_n_8886142.html.

Yuan, X.C., Wang, Q., Wang, K., Wang, B., Jin, J.L. & Wei, Y.M. 2015. China's Regional Vulnerability to Drought and Its Mitigation Strategies under Climate Change: Data Envelopment Analysis and Analytic Hierarchy Process Integrated Approach. *Mitigation and Adaptation Strategies for Global Change* 20: 341–59.

Yun, S.-J., Ku, D., Park, N.-B. & Han, J. 2012. A Comparative Analysis of South Korean Newspaper Coverage on Climate Change: Focusing on Conservative, Progressive, and Economic Newspapers. *Development and Society* 41(2): 201–28.

Yuval-Davis N. 2011. *The Politics of Belonging: Intersectional Contestations.* Thousand Oaks, CA; New Delhi; Singapore: Sage.

Zalasiewicz, J. 2015. The Geology behind the Anthropocene. Typescript.

Ziervogel, G., Pelling, M., Cartwright, A., Chu, E., Deshpande, T., Harris, L., Hyams, K., Kaunda, J., Klaus, B., Michael, K., Pasquini, L., Pharoah, R., Rodina, L., Scott, D. & Zweig, P. 2017. Inserting Rights and Justice into Urban Resilience: A Focus on Everyday Risk. *Environment and Urbanization* 29(1): 123–38.

Zinn, H. 2004. The Optimism of Uncertainty. *The Nation.* September 2, 2004. www.thenation.com/article/optimism-uncertainty/.

Zuhair, M.H. & Kurian, P.A. 2016. Socio-economic and Political Barriers to Public Participation in EIA: Implications for Sustainable Development in the Maldives. *Impact Assessment and Project Appraisal* 34(2): 129–42.

INDEX

231, 243; South 2, 7, 12, 136, 253–5; *see also* individual countries
American Enterprise Institute 122
American Geophysical Union 58
Americans 95, 117–18, 127–8, 265–6; African 100, 139; Arab 100; Mexican 100; native 112–13, 100; *see also* United States
anarchists 47
ancestry 70–2, 80
Andersen, Hans Christian 62
Andes 253–60
animal: emissions xxii; rights 27
animals 1, 6, 14–15, 17, 23, 27–8, 33, 48, 70, 76, 83–5, 87–8, 93, 100, 105, 141–3, 164, 168, 256, 261; human attitudes towards 100 (*see also* human–nonhuman connections); marine 23
Anjiara, Jonathan Shapiro 77
Anthropocene era 4, 24–5, 27–8, 30, 34, 57–8, 60, 79, 108, 110, 149
Anthropocene Working Group 60
anthropocentrism 27, 61, 149, 257–8, 261
anthropologists 253
anti-capitalism 113
anti-colonialism 20, 111
anti-racism 142
apartheid 133, 135
apes 93
apologies 80
Appadurai, Anjali xi
Apple Inc. 133
Applied Research Center 98–9
aquifers 71
Arab-Americans 100
Archer, David 23, 30
Arctic 108, 216–17, 226, 267; Arctic Circle 267
Argentina 130
Ariafa, Arison 167
ariel (journal) 34
Arizona 99, 114
Arkansas 99
army, British: Pioneer Battalion 170; US 92
Arora-Jonsson, Seema xi
Arquero, Darren 98
arsenic 141
art, indigenous 85
artefacts, museum 170–1
articles, academic journal 195

artists 111
"artivism" 229
Arvidsjaur (Sweden) 91n1
Åsele (Sweden) 88
Asia 7, 216; Central 195, 198n5; Southeast 195, 198n5; Western 198n5; *see also* China; India; Japan; Korea; Pakistan; Vietnam
Asian Development Bank 156
Asians 100; South, in Britain 74
Assadourian, Erik 64
assets, capital 51
assimilation 174
assistance, technical 185
assault, sexual 97
Atal Mission for Rejuvenation and Urban Transformation (AMRUT) 202, 211
Atkinson, Joshua 124
atmosphere xxii–iv, xxvii, 22, 24, 36–9, 41, 44, 107, 115, 141, 143, 192, 220, 252
atmospheric space xxiii, xxvi–vii, 22, 220
atolls 155–6, 160, 167n1, 168–9, 175, 182n1
Atolls Integrated Development Policy (PNG) 161
Atwood, Margaret 1–2
Auckland (NZ), airport 168
Audre Lorde Project 98
Aufdringlichkeit (obtrusiveness) 119
Aufsässigkeit (obstinacy) 119
austerity 52
Australia xxiii, 26, 30, 122–3, 125, 156, 170–2, 238, 243; Western 66
The Australian (newspaper) 122
Austronesia 15
authoritarianism 107, 143, 180
autocracy 176–7
automobiles *see* cars
aviation, commercial 168–9, 265
awareness, community 163–5; environmental 165
Axelsson, Per 84
Aymara people 253, 256–7
Ayni (reciprocity/mutualism) 258
Aznar, José María 237

Baber, Walter F. xi, 225
Baca, Krystal 46
Baconianism 61

ZED

Zed is a platform for marginalised voices across the globe.

It is the world's largest publishing collective and a world leading example of alternative, non-hierarchical business practice.

It has no CEO, no MD and no bosses and is owned and managed by its workers who are all on equal pay.

It makes its content available in as many languages as possible.

It publishes content critical of oppressive power structures and regimes.

It publishes content that changes its readers' thinking.

It publishes content that other publishers won't and that the establishment finds threatening.

It has been subject to repeated acts of censorship by states and corporations.

It fights all forms of censorship.

It is financially and ideologically independent of any party, corporation, state or individual.

Its books are shared all over the world.

www.zedbooks.net
@ZedBooks